Understanding Anxiety, Depression and Bipolar Disorder

A New Theory of the Six Primary Emotions

Understanding Anxiety, Depression and Bipolar Disorder

A New Theory of the Six Primary Emotions

William W. Hedrick, M.D.

ISBN 978-0-578-07905-9

To

I dedicate this book to Merrie, my wife and best friend, who has stuck by me through thick and thin, and to my children Robert, Peter, Ben, Katherine, William, Matthew, Sallie and Jonathan, and to my good friend Dr. Nicholas Stratas who encouraged me to pursue this project and gave me hope that it would succeed.

Preface

I am a family physician, not a psychiatrist, but when I went into practice in the early 1960s I quickly realized that a great deal of my practice was going to be devoted to patients with psychiatric problems. Even though I had had almost a year of training in psychiatry, I still found myself struggling to deal with the many emotional problems that patients had. When dealing with medical illnesses there were discernable causes and normal levels I could use to determine when and why someone was sick, but with emotional illnesses there were neither causes nor normal levels to offer any guidelines to help me determine when someone had an illness. It was like a guessing game. Sometimes when I tried to fit a patient's symptoms into what I had been taught, it was like trying to fit a square peg into a round hole.

Dr. Freud said that depression was anger turned inward, but most of the patients I saw with depression showed no signs of anger. I saw wives who complained of anxiety, but when I got their history I found they were being abused by their husbands, or had been abused as children. So, was their anxiety normal? I saw patients who had symptoms of depression, anger, and anxiety, sometimes all three at the same time, but when I turned to books on psychiatry and psychology for answers, there was nothing in the books that said that a patient could have all three at the same time. It was all very confusing. What was in the books didn't match what I was seeing in the real world.

There was one thing missing from the books. How is someone supposed to feel? What is normal? I felt that if I just had some guidelines about what is normal, it would be easier to say what was abnormal. If someone has a loved one to die, how are they supposed to feel? Are they supposed to feel depressed, or sad? Which emotion is normal? Is there a difference between depression and sadness, or are they the same thing? How about anxiety? We all get anxious sometimes, but how much anxiety is normal? I couldn't find an answer to any of these questions in the disciplines of either psychology or psychiatry.

How to define "normal" was the elephant in the room. As my frustration grew I began a search for a paradigm that would explain what I was seeing in the real world. I found that not being classically trained in psychiatry has been in some ways an advantage, because when I began my search I had no prejudice toward any particular school of psychiatry or psychology. Since there were so many things that didn't fit into the conventional wisdom, it became apparent that it was going to take a new paradigm to explain what I was seeing in the real world, one that was different from what I had been taught. The term "chemical imbalance" is often used to explain mental illness, but the concept of a chemical imbalance has changed little since Hippocrates tried to explain depression more than 2000 years ago as a chemical imbalance, except today black bile has been replaced by serotonin.

Some of the terms used in psychiatry and psychology are often confusing. For example, is depression a mental illness, or an emotional illness? What does it mean to be mentally ill? What does it mean to be emotionally ill? Are they the same, or are they different? We use both terms, but there is no general agreement about what each means. Popular diagnoses come and go. In the 1800s women were affected by the vapors. In the 1950s the diagnosis of conversion hysteria was all the rage. At one time homosexuality was included in the list of psychiatric disorders. Today the diagnosis of bipolar disease is very trendy, and almost every teenager who acts out or is somewhat belligerent is tagged as having bipolar disease. A look at the history of psychiatric diagnosis show that diagnosis have often come and gone like epidemics.

It became apparent that in order to accomplish my goal I needed a paradigm that reflected two elements not mentioned in the current lexicon. The first was a rational cause for mental illness, one that explained things in terms of something other than "chemical imbalances," and the second was a definition of normal. Most mental health providers today use the *Diagnostic and Statistical Manual of Mental Disorders* as guide and gospel. It is published by the American Psychiatric Association, and its foundations go back to Dr. Emil Kraeplin and his work in the late 1800s and early 1900s. In the DSM the diagnosis of a particular mental disorder, (the term "mental illness" is never used in the manual, only "mental disorders"), is based on a check-list of symptoms. If a patient has a given number of symptoms listed for a given disorder, by DSM criteria they qualify to have the "disorder". The only reference to a normal emotion in the book is the mention of normal bereavement in the loss of a loved one and the only cause listed for any of the psychiatric disorder is a passing reference to a "chemical imbalance" in the case of depression.

The approach that the DSM takes toward psychiatric illness is based primarily on whether an individual can function in society, and/or whether their symptoms are causing them discomfort. Nowhere in the manual is the difference between a normal emotion and an abnormal one ever made clear. Psychiatric theory, from Dr. Freud on, has always begged the question when it came to defining normal, but without some reference to normal, it becomes very difficult to say what is abnormal. The old expression "if it walks like a duck and talks like a duck, it must be a duck," is no longer adequate for psychiatric diagnosis.

When I was in training as an intern, and later as a resident, we were taught to measure everything we could about the physical aspects of the patient: the specific gravity of the urine, the amount of hemoglobin in a cc of blood, x-rays, the white count, the blood pressure, the amount of sodium and potassium in the blood, and so on. We made these measurements in order to determine whether the patient had a result outside the normal limits and to be able to say whether the patient was sick or not. We would also use the data to help us make a diagnosis. When we got to psychiatry there was nothing to measure, and never a reference to normal. This was especially true when we studied Freud's superego, ego, and id. I felt that if I could find something about the emotions that I could measure, like we do the blood sugar and cholesterol, it would be possible to establish normal limits for the emotions, and allow me to diagnose and treat patients with emotional illnesses the same way I did medical illnesses.

Over the past forty-seven years my office has been my laboratory, and my patients have been my guide as I searched for a rational approach to mental illness. During those years I have kept what worked and made sense, and discarded what didn't. This book is about my search for a paradigm that explains mental illness from a real world perspective, one that is reasonable and rational, and one that puts mental illness and medical illness on the same plane; but as the reader will discover, it often runs counter to what is currently being taught.

The cartoons in the book were drawn for me by Rowell Gorman, a local artist, for a series of lectures I gave in the 1980s. I would like to thank my first cousin's wife, Molly Weston, for helping me edit the book.

Table of Contents

1. Preface vii

2. Foreword xv

3. A Brief History of Mental Illness and Drug Treatment 1

   Dr. Emil Kraepelin 5

   Some Thoughts on Schizophrenia 6

   Dr. Kraepelin's Goal 9

   Dr. Sigmund Freud 9

   Psychoanalysis 11

   The Decline of Psychoanalysis 13

   The DSM Series 14

4. Drugs and Other Treatments 17

   The Fever Cures 18

   The Early Drugs 20

   The Coma and Shock Therapies 21

   The Discovery of the Neurotransmitters 24

   The First Effective Psychotropic Drugs 27

   The First Specific Anti-anxiety Drugs 30

   The SSRI's and Beyond 33

5. Medical School, Internship and an Unexpected Year in
   Psychiatry 37

   A Year of Psychiatry 42

   The U.S. Army and France 45

   Two Years in France 47

   Home From France and Back to Grady 54

   Returning Home and Starting a Practice 57

6. The Search for a Solution to a Vexing Problem.............................. 61

Dr. Ellis' 10 Irrational Ideas. 62

Arriving at a Definition of Normal 64

Different Theories on the Emotions...................................... 67

Redefining the Primary Emotions.. 73

A Redefining of the Primary Emotions 74

The Six Primary Emotions.. 76

7. Anxiety, the Emotion of Escape from Harm and a
Model for the Other Primary Emotions................................. 79

Panic Attacks and Post Traumatic Stress Disorder...................... 86

8. A New General Theory of the Primary Survival
Emotions.. 89

Making a Distinction between Mental and Emotional
Illness 92

Bipolar Disease as an Example of Emotional Center
Malfunction... 93

Tripolar and Quadripolar Disease.. 94

Anxiety Center Malfunction 95

Some Additional Thoughts on Agoraphobia and Post
Traumatic Stress Disorder.. 98

The Neurotransmitters of the Primary Survival
Emotional Centers.. 100

The Master Emotional Neurotransmitters.............................. 102

9. Anger-Aggression, the Emotion of Possession and
Control... 105

Evil and Anger-Aggression... 109

10. Depression, the Emotion of Despair and Hopelessness 115

11. Manic, the Emotion of Self-Confidence and Hope 125

Some Additional Notes on Bipolar Disease............................ 127

12. Sadness, the Emotion of Loss 131

13. Happiness, the Emotion of Gain 137

14. Some Notes on Causes of Malfunction in the Four
    Primary Survival Emotional Centers 141

15. Some Additional Notes on Why Patients Are Often
    Confused About Which Emotion They Are Actually
    Feeling 143

16. Some Additional Thoughts on Freud's Psychology and
    Addiction Disease 145

    Addiction Disease 148

    Endorphin Addictions 149

    Nicotine Addiction 153

    Addiction to Marijuana and the other Hallucinogens 154

    Alcoholism, Another Form of Endorphin Addiction 154

    Runner's High 155

    Addiction to Pain 155

    Addiction to the Manic Neurotransmitter 156

    Happiness Addiction 158

    Sexual Addiction 159

    Thrill Addiction, the Addiction to Danger 161

    Treatment of the Different Addictions 161

17. Some Final Notes on Programming, Cognition, and
    Relationships 163

18. The Difference between Remorse and Guilt 167

19. One Final Note 169

    Post Script: 173

20. References 175

Foreword

Where do our feelings and emotions come from? Do they arise from inside of us, or come from outside of us? For centuries man felt it was the gods or the spirits that put emotions into our bodies and thoughts into our minds. When Homer wrote the Iliad and Odyssey many centuries ago, his mortals were at the mercy of the gods, and under their control; and even though we might think we have come a long way since then, the idea that our thoughts and feelings are controlled by outside forces has not entirely disappeared. There are some who still believe that evil comes from some external force which causes someone to do evil things. Satan gets the blame for much of the evil in this world, and even today some church figures feel that mental illness is caused by satanic possession. This attitude is illustrated tragically by an incident which occurred in 2005 in a monastery in Romania. When one of the nuns developed hallucinations and delusions from schizophrenia, the head priest decided that her illness was due to satanic possession and ordered her treated with exorcism. He had the nuns of his order chain her to a cross and stuff a towel in her mouth to keep her from screaming. After three days on the cross she became so weak some of the nuns became alarmed and called an ambulance. Tragically, she was dead by the time she got to the hospital. An autopsy showed she had died of malnutrition and dehydration.

Today, we know through modern research that our emotions and feelings do not come from the outside, but instead come from the neurotransmitters that course through our brains and bodies; and, even though we know a great deal about those neurotransmitters, there is still much to be learned. What are the primary emotions? Where do the emotions come from, and what actually causes us to feel an emotion? What distinguishes one emotion from another? Is it possible to measure an emotion and be able to say how much of a given emotion is normal, or to say when the amount of an emotion falls outside the normal limits like we can with the blood sugar in diabetes?

I once sat on the mental health committee of our state medical society, and during one of our meetings the subject came up as to why mental illness is such a problem for patients, the public, doctors, industry, politicians, and the health insurance companies. During the discussion, one of the members spoke up and said that he couldn't understand why there was so much misunderstanding when we have the DSM, the guidebook published by the American Psychiatric Association which classifies the different mental disorders. The member sitting next to me turned and whispered, "Yeah, but it doesn't tell you what normal is." And I believe one of the primary reasons why there is so much confusion about mental illness is because psychiatry has never been able to define normal, and because it has never been able to normal, defining abnormal remains problematic.

There are many things about the brain and body that we understand, and many things about the brain and the body we can measure, but when it comes to the emotions we generally are at a loss. We can measure the blood sugar, thyroid function, kidney function and so on. We know what the normal range is for these tests, and we know that if a result falls outside the normal range, it is because of an illness. If a patient has a blood sugar of 500, we know that this is abnormal, and the patient likely has diabetes. We can measure brain waves, get MRI scans of the brain, run psychological tests, but the question remains, how can we measure an emotion? Is it possible to say when an emotion is normal or abnormal?

Mental illness has always been something of an anathema to the general public, and no less a problem for Congress and the managed care industry. Today if an illness is labeled mental or emotional, reimbursement for treatment by Medicare is automatically cut by 50 percent, (efforts are currently underway to correct this), despite the fact that recent studies have shown that depression carries with it as much morbidity and mortality as coronary heart disease. Why is it that society and the health care industry consider one illness as being acceptable and the other not?

With medical illnesses we have parameters that we can measure and this makes it easier to determine when something falls outside of a normal range. We can measure the blood pressure and compare it to a known standard. We can follow its progress, or lack of progress with measurements. We have goals and guidelines for the blood pressure, but what about an emotional illness like depression? You can't draw a blood test to rule it in or out. You can't hook the patient up to a monitor and get a diagnosis. We can talk to the patient and get a history. We can observe their behavior and symptoms and get an estimate of how well they are functioning, but we still don't have anything that we can measure or show an insurance company.

In medical journals we are constantly reminded that depression is not being diagnosed or treated nearly as often as it should be. And depression is not the only emotional problem that is under treated. Studies show that anxiety disorders are also extremely common, and we know that when both of these illnesses are treated patients can lead normal and productive lives. Doctors are constantly bombarded with information about how to diagnosis and treat this or that emotional disorder, yet countless thousands of people go untreated. Why is this? One of the major deterrents to the treatment of mental illness is the longtime stigma attached to it by both laypersons and professionals, and I believe part of the blame for this dilemma lies with the current paradigm as epitomized by the DSM series. As my colleague pointed out in the mental health meeting, the DSM doesn't tell you what normal is. And, until we have a paradigm that includes a definition of normal, we are destined to remain in the dark. Edward Shorter, author of *The Story of Psychiatry*, in critiquing the DSM, points out that diagnoses over the years have been added or taken from the series, not on the basis of scientific evidence, but often because of political pressure. He goes on to say, "The underlying failure to let science point the way emphasized the extent to which DSM-lll and its successors, designed to lead psychiatry from the swamp of psychoanalysis, was in fact guiding it into the wilderness."

The basic premise of the DSM is that a mental or emotional illness can be diagnosed in the same way a cook uses a recipe except in reverse. If all of the ingredients of a cake are present, then it must be a cake, (except in the DSM you need only to have a certain number of the ingredients). For each diagnosis there is a list of symptoms associated with the disorder, and if the patient has a given number of the symptoms associated with a given disorder, (in the case of anxiety four out of ten), by DSM criteria the patient has the disorder. In an effort to make it easier to diagnose a given disorder, the book leaves itself open to much individual interpretation, and studies have shown that if a patient is seen by several psychiatrists, it is not unusual for them to disagree on the diagnosis and some diagnoses have a tendency to be over diagnosed such as bipolar disease.

In their book *The Loss of Sadness, How Psychiatry Transformed Normal Sorrow Into Depressive Disorder*, Allan V. Horwitz and Jerome C. Wakefield point out one of the major weaknesses of the DSM series, which is by relying only on symptoms, it becomes very easy to confuse different psychiatric diagnosis, such as sadness and depression. Another book very critical of the DSM series is *The Selling of DSM-The Rhetoric of Science in Psychiatry*, by Stuart A. Kirk and Herb Kutchins. In the frontispiece of their book they have a wonderful quote from *Alice in Wonderland* about the DSM, which also says something about the problem:

"What's the use of their having names," the Gnat said, "if they won't answer to them?" "No use to them," said Alice, "but it's useful to the people that name them, I suppose."

Despite the best intentions of the authors of the DSM, its primary weakness lies in the fact that it is essentially groundless. Except for a vague reference to a chemical imbalance in the case of depression, no cause for any disorder is ever proposed. Normally, in any kind of nosology, or classification, there exists a cause for what is being classified, or at least a reference to some standard. The DSM lists hundreds of different disorders, but makes no reference to a theory that adequately explains the cause of any psychiatric illness or disorder. It is as if mental illness were created out of whole cloth.

The DSM often comes across as being very vague. For instance, the DSM-IVR defines a panic attack as: *a discrete period of intense fear or discomfort, in which four (or more) of the following symptoms developed abruptly and reached a peak within 10 minutes:*

The symptoms:

(1)   palpitations, pounding heart, or accelerated heart rate
(2)   sweating
(3)   trembling or shaking
(4)   sensations of shortness of breath or smothering
(5)   feeling of choking
(6)   chest pain or discomfort

(7)   nausea or abdominal distress
(8)   feeling dizzy, unsteady, lightheaded, or faint
(9)   derealization (feelings of unreality) or depersonalization (being detached from one's self)
(10) fear of losing control or going crazy
(11) fear of dying
(12) paresthesias (numbness or tingling sensations)
(13) chills or hot flushes

As a physician, if I saw a patient with a combination of four or more of these symptoms, I wouldn't think of a panic attack, I would think that the patient had some kind of heart or lung disease. To make a diagnosis of panic attack based on these symptoms alone would, in my opinion, be foolhardy. A major medical illness would need to be ruled out, and in reality most of the time when the diagnosis of a panic attack is made, it is usually made after the patient has been thoroughly checked out and no medical illness found, meaning that the diagnosis of a panic attack is almost always made as a diagnosis of exclusion.

Another unrealistic and confusing criterion, which is often listed in the DSM, is in order to make a diagnosis the patient must have the symptoms for a specific length of time. In the above example of a panic attack, the symptoms must peak after ten minutes. If the symptoms peak after five minutes by DSM criteria, one can't make the diagnosis. In another example of a time-related criterion, that is equally absurd; it states that after the death of a loved one:

The diagnosis of Major Depressive Disorder is generally not given unless the symptoms [sadness] are still present 2 months after the loss.

It does say that the duration and expression of "normal" bereavement can vary considerably, but it seems to imply that sadness and depression are basically the same, and at one point it states that a sad mood is one of the symptoms of depression, but it never makes a clear distinction between depression and sadness, and implies that depression is always abnormal. Contrary to the DSM, I believe depression was built in to help our ancestors survive helpless, hopeless circumstances and is a primitive form of hibernation, and even today is normal when someone is truly helpless and hopeless.

Today there is a great deal of research being conducted on the emotions and neurotransmitters, and our knowledge of the connection between the emotions and neurotransmitters is growing rapidly. In Dr. Candace Pert's book, *The Molecules of Emotion*, which was published in 1999, she makes the case that everything we think or feel is basically chemical. When we feel happy, it is because we have happy chemicals in our brains and bodies. When we feel angry it is because we have anger chemicals in our brains and bodies. When we feel anything it is because we have in our brains and bodies the neurochemicals that make us feel that way. We feel an emotion the same way we feel anything else. We feel a hot stove when the nerve endings in our fingers relay to the brain the sensations of heat and pain from the heat and pain sensors in the fingers. Likewise, when we feel an emotion it is because the brain is registering the neurotransmitters of that emotion. Emotions and feelings don't just come out of nowhere. There is also much confusion about which emotion is which, and one of the reasons this is the case is because our emotional neurotransmitters don't carry any labels with them. When I am counseling a patient in the office, I can never be sure that the patient and I are talking about the same thing. Different patients use the same words to describe entirely different feelings and emotions.

One of the major problems contributing to the confusion about the emotions is that it has never been settled among philosophers, psychologists, or scientists where the emotions come from, or what they represent; nor has it been settled which ones are the primary emotions. Another issue contributing to the confusion about the emotions is a lack of agreement on whether there is a difference between feelings and emotions. In his book on Spinoza and the brain, *Looking for Spinoza*, Dr. Antonio Damasio says that he believes that feelings are of a higher order than the emotions, but I see no reason to make such a distinction. When we feel anything, whether it is physical, such as heat or cold, or an emotion, such as anxiety, it is because the brain is registering the neurotransmitters of that particular feeling and to say that one is higher than the other is, in my opinion, purely arbitrary.

Today we know a great deal about the neurotransmitters and hormones that maintain the internal environment of our bodies, including those that regulate blood sugar, blood pressure, thyroid, and so on. What we don't talk as much about, and will be the primary focus of this book, are the neurotransmitters that enable us to survive in and relate to the external environment, which I believe are the primary emotions. We know what the normal levels are for the chemistries of the internal environment, and when they are abnormal. The purpose of this book is to propose a paradigm that treats the neurotransmitters of the external environment, the primary emotions, the same way we treat the chemistries of the internal environment, and to be able to say when they are normal, and when they are abnormal.

When I first went into practice in the early 1960s, one of my primary goals was to try to take care of the whole person, which meant not only treating their physical problems, but their emotional problems as well. I felt comfortable with their medical problems, but I wasn't so comfortable with their emotional problems, even though I spent almost a year as a psychiatric resident at Dorothea Dix Hospital in Raleigh, and had been chief of the psychiatric unit for a time at the 34th General Hospital in France during my Army tour.

When I first went into practice in the early 1960s, we had psychoanalysis and a few drugs. Psychoanalysis wasn't very practical for the family doctor's office, and most family physicians referred patients to psychiatrist. In the late 1960s and early 1970s a revolution in the treatment of mental illness took place with the introduction of new psychotherapies and more effective drugs. Studies at the time showed that the new drugs and the new psychotherapies were often equally effective in the treatment of minor depression and some cases of anxiety, and this brought to the fore a question nuch debated in psychiatric circles. Is psychiatric illness a psychological problem, or a chemical problem, or both? Almost from the very beginning of psychiatry there has been a split between those who feel that psychiatric illness is primarily a psychological problem, with treatment being some type of psychotherapy; and those who feel that psychiatric illness is caused by some kind of chemical imbalance, with a biological cause, and treatment being medication. To date there is no good theory that successfully integrates these two schools of thought and explains how both psychotherapy and drug therapy can both work in the same illness. One of the aims of this book is to integrate these two schools and offer an explanation for how drugs and psychotherapy can both work in the same illness.

I believe that if we are ever going to destigmatize mental illness, it is essential that we arrive at a definition of normal. I remember when I was a psychiatric resident I kept bringing up the question, "what's normal and what are our goals in treatment?", but no one seemed to have a good answer. The answer I was almost always given referred to the patient's ability to function. If the patients' symptoms interfered with their ability to function, it meant that they had a mental disorder. This answer always seemed very vague to me because it lacked a theoretical basis. Ability to function, compared to what and when? I once posed a similar question at a conference on anxiety at UNC in Chapel Hill, N.C. I asked one of the moderators, "What is the difference between normal anxiety and abnormal anxiety?", and I received essentially the same answer. "It's abnormal," he said, "if it interferes with the patient's ability to function." Once again I didn't feel his response answered the question, and that his answer essentially begged the question.

Mental illness is a problem that affects so many of our lives in one way or another, sometimes directly, sometimes indirectly, and there has always been a great deal of mystery and fear surrounding it. But what if we understood it the same way we understood diabetes, or high blood pressure, or thyroid disease? Maybe then we wouldn't be so afraid of it.

I have been a family physician for more than forty years and this has given me the opportunity to observe patients and their families from a unique perspective. Some of the patients who were in their twenties when I went into practice, are now on Medicare. I have treated the parents, the children, the grandparents, the grandchildren, and even the great grandchildren. I have seen firsthand how both genetics and the environment can affect individuals and families, how both nurture and nature operate in the human species.

As a family physician I have tried to offer my patients a continuity of care over time, and the treatment of the whole person. Most patients don't like to see one physician this month, and another the next. They want a physician they can call their own and be someone they can depend on. The latest term for this concept is called a medical home. But today so much of health care is delivered through a system called managed care, which is often fragmented and not very people-friendly. It is a system where people can be treated like objects, and their care manipulated to the advantage of the company, and where the ultimate goal of the managed care companies, in my opinion, is not what is best for the patient, but what is best for the company's bottom line.

In their book, *Critical Condition, How Health Care in America Became Big Business and Bad Medicin*e, Donald L. Barlett and James B. Steele tell the story of how this came about in the 1970s and 1980s when a group of MBAs decided that introducing the business model into medical care could be profitable for themselves and the companies they created. Huge companies took over how most medical care is delivered, and what has happened since then, in my opinion, has been a disaster. The attitude has changed from one of caring, to one in which the patient might as well be a car that needs repair or maintenance, and sometimes their attitude would seem to be the car might as well be junked. Managed care has produced a climate in which the treatment of illness is minimized, so that profits can be maximized for the share holders, and where the CEOs who run these companies can line their pockets with outrageous salaries. It is not surprising that so many patients and physicians are frustrated and unhappy with the climate of today's medical care.

Because I see patients with both medical and emotional problems, it is much easier for me to see the patient as whole person and not just as an object with broken parts. I sometimes feel that the managed care companies would have us believe that patients could just as easily be treated by a computer, but most of the patients I see have such a unique set of problems that no computer could ever unravel them. On a daily basis I get letters and faxes from the managed care companies telling me how treat patients, based on the latest buzz word, *evidence-based medicine*. They also try to tell me which drugs to use, but I know that the ones they recommend are the ones that they can buy the cheapest, and may not necessarily be the best drug for the patient.

Sometimes when I want to get an MRI, I have to call and speak to a clerk who appears to have a check-off list in hand. It is not unusual for that person to have an accent that leads me to believe they are on the other side of the earth, and I have to try and convince that person to let me order an MRI on a patient they have never seen.

Sometimes I feel like a pawn in the service of the managed care companies, which leads to an adversarial relationship and one which fosters an atmosphere of resentment among both patients and physicians. Patients don't want to feel that their illnesses and lives are a commodity that can be used to make a profit for a managed care company. They want to be treated like individuals and physicians don't want to feel like cogs in some gigantic health care machine, whose primary purpose is to churn out profits for the company. I know there has to be a business side to medicine, but I think that it is past time for the pendulum to swing back toward a concern for the patient, and away from the company's bottom line.

I believe that each patient who comes to my office is a unique individual who deserves to be treated with the same respect and care as every other individual, no matter the person's education, wealth, or place in society. I have patients from all walks of life, lawyers from the wealthy neighborhood near my office, and tenant farmers who live in houses without plumbing. I treat them all the same.

I also try to keep this same attitude toward whatever illness the patient has, whether the illness is diabetes or depression. I try to treat a patient with drug addiction the same way that I treat a patient with hypothyroidism. Unfortunately this attitude of equality is not always the case in society, or with the powers that be. All too often society looks at medical illness and emotional illness very differently. Individuals with an emotional illness are sometimes made to feel guilty because they suffer from anxiety or depression, and made to feel that it's their fault because they have an emotional problem, and if they were just strong enough they should be able to do something about it on their own. To me that is like blaming a patient with hyperthyroidism because their thyroid gland puts out too much thyroid hormone, or blaming a diabetic because their pancreas gland doesn't put out enough insulin. Over the last few years the attitude toward mental illness does appear to be getting better, albeit slowly and today drugs like Zoloft for depression, and the "purple pill" for acid reflux, are being pitched side by side on television.

For far too long emotional illness has been tagged as being "all in the head," but I think the time has long since passed when that phrase should be retired as a cover-up of our ignorance. I also think that the phrase "chemical imbalance" is overly simplistic and doesn't say or explain very much. Today we have the means to treat the vast majority of mental illnesses with the same ease we treat high blood pressure, but tragically the number of people who get treated for mental illness is still very low. One of the aims of this book is to help correct that inequity.

I would like to begin the book with a brief history of mental illness and how treatment and theory have changed over the years, then a section on the development of psychiatric drugs, followed by a brief section on my own training and background, and finally my search and arrival at a paradigm that explains mental and emotional illness from a more rational perspective.

A Brief History of Mental Illness and Drug Treatment

References to mental illness go back as far as recorded history. The Bible, for example, contains a number of instances of mental illness. One example is from the Old Testament in the First Book of Samuel, Chapter 16, from the story of King Saul and David:

The Lord's spirit left Saul and an evil spirit sent by the Lord tormented him. His servant said to him 'We know that an evil spirit sent by God is tormenting you. So give us the order, sir, and we will look for a man who knows how to play the harp. Then when the evil spirit comes on you, the man can play his harp, and you will be all right again.

"Saul ordered them, 'Find me a man who plays well and bring him to me.......'

"David came to Saul and entered into his service......From then on whenever the evil spirit sent by God came on Saul, David would get his harp and play it. The evil spirit would leave, and Saul would feel better and be all right again.

The book of Samuel says that the evil spirits that tormented King Saul would come and go, and even though manic episodes are not mentioned, in all likelihood he suffered from what is known today as bipolar disease. The only treatment they knew of at the time for the depressive phase of his illness was music therapy, and if he had manic episodes, they probably weren't recognized, since he was the King.

Another example in the Bible is from the New Testament. A man is brought to Jesus with some form of madness, and Jesus cures the man by transferring the evil spirits into a group of pigs. The Bible goes on to say,

....about two thousand [pigs] in all, rushed down the side of the cliff into a lake and were drowned, taking the evil spirits with them, and leaving the man cured.

In both of the Biblical illustrations, mental illness is depicted as something that enters the body from the outside, because in ancient times, (and for some this is true even today), it was felt that madness was something that someone caught like an infection, or came from a spell that the gods cast upon someone; and the ancients felt that the only way to get rid of the evil spirits was through exorcism, or some type of cleansing rite. So purging, bleeding, and laxatives were often used in an effort to rid individuals of their madness.

The first physician to propose that mental illness came from natural causes, and not supernatural ones, was Hippocrates who lived from 470-377 BC. Hippocrates had the very modern notion, which was contrary to most of his contemporaries, that mental illness came from something that went wrong in the brain and in one of his books he wrote the following:

....*the brain is the sole origin of pleasures and joys, laughter and jest, sadness and worry, as well as dysphoria and crying. Through the brain we can think, see, hear, and differentiate between feeling ashamed, good, bad, happy... .Through the brain we become insane, enraged, we develop anxiety and fears, which can come in the night or during the day, we suffer from sleeplessness, we make mistakes and unfound worries, we lose the ability to recognize reality, we become apathetic and we cannot participate in social life... . We suffer all of those mentioned above through the brain when it is ill... .*

And lest we think that the term "chemical imbalance" is a new concept, Hippocrates theorized that mental illness came from an imbalance in one of the body's four humors; blood, black bile, yellow bile and phlegm. What we call depression, he called dysphoria, and he felt that dysphoria came from too much black (melan), bile (cholia), from which we get the word melancholia.

Today we know that most mental illnesses come from something that goes wrong inside the brain, but it is well to remember that there are still some mental illnesses that do come from the outside, and not from satanic possession, or a curse the gods put on someone. Certain infections can cause mental illness. There was a time when one of the most common forms of mental illness was tertiary syphilis, an infection which causes psychosis in its late stages. In the nineteenth century syphilis was very common, and at the time it was impossible for a physician to tell the difference between mental illness caused by syphilis and mental illness from some other cause. There were a number of well-know historical figures in the nineteenth century who became insane in their latter years due to tertiary syphilis, including Robert Schumann, the composer, Friedrich Nietzsche, the philosopher, and Eduard Manet, the painter.

Today there is an infectious agent that causes mental illness and dementia that is even bigger than syphilis was it its day. AIDS, an infection caused by HIV, can cause dementia and hallucinations in its late stages.

We need to remember that it was only a few centuries ago when people with severe mental illness were confined to asylums, where they were sometimes chained to a wall until death mercifully relieved them of their misery. Occasionally they were kept more humanely in private homes, but almost always away from the public. They were often thought of as being more like animals, and were frequently treated as such. As late as the early 1800s at the Pennsylvania Hospital for the Insane in Philadelphia there was a section where the public could pay admission to view the "lunatics" who were exhibited in cages along the street like exotic animals in a zoo.

In the late 1700s a revolution took place in the way people saw themselves in relationship to the world and to each other. It was called the Age of Enlightenment and the worth of the individual suddenly gained new importance and from this new perspective there arose a sense of optimism that all of man's problems could be solved through science and reason. In medicine the doctors who treated the mentally ill were caught up in this new spirit of optimism, and the asylums were changed almost overnight from prisons to hospitals. This change in attitude also gave rise to the hope that for the first time in the history of medicine; patients with severe mental illness could be helped, and as this new attitude took hold doctors began treating patients with mental illness more like human beings.

One of the first physicians to propose that asylums could become places of healing was an English physician by the name of William Battie. In 1754 he wrote a *Treatise on Madness,* in which he put forth the revolutionary idea at the time, that individuals with mental illness should be treated the same way as individuals with physical illnesses. Across the English Channel, in France, Dr. Philippe Pinel became that country's most well-known reformer. It was Dr. Pinel who first ordered the removal of the chains from the patients at Salpetiere Hospital in Paris. More than anyone else it was Dr. Pinel who embodied this new spirit of optimism. He felt that the asylums could become places of healing, and was one of the first to offer hope for society's lost souls.

Doctors who treated the mentally ill up until the early 1800s were called alienist, presumably because they were treating patients whose minds appeared to be alienated from the body and reality. Sometimes they were called asylum doctors. A physician by the name of Johann Reil, who was a professor of internal medicine and neuroanatomy at the University of Halle in Germany, was the first person to refer to physicians who treated the mentally ill as psychiatrists. The word has its origins in the Greek word "psyche," which means the soul or mind, and "artiera," which means to heal or cure.

Up until the 18th century very little attention had been paid to the minor psychiatric illnesses, but with the new spirit of optimism that accompanied the Age of Enlightenment, there came hope that even those with minor problems, like anxiety and hysteria, could be helped. Women in the 1700s were often affected by a condition called the "vapours." An attack of the vapours was usually brought on by some kind of stressful social situation, and when a woman was attacked by the vapours she would throw her arm across her face and slump unconscious to the floor. Smelling salts usually was effective in reviving the poor lady, and most women carried a small bottle in their purse to treat this mysterious illness.

With the Enlightenment there also came a great deal of curiosity about how the brain and body work. This led to extensive research during the 18th and 19th centuries by some of the early pioneers in medicine on the brain and the nervous system. It was felt that the "vapours," and some of the other nervous disorders, like hysteria, were caused by abnormalities of the nervous system, and these conditions were attributed to "bad nerves."

As it became more acceptable to have one of the less severe nervous conditions, there arose a group of physicians who specialized in their treatment. Up until this time the family doctor had usually been the one who treated patients with minor psychiatric conditions, but in upper-crust society, especially in Europe, patients began seeking the help of a "society" physician, someone who specialized in the treatment of "nerves" in their office. The "society" physician became the forerunner of today's office-based psychiatrist.

Dr. Benjamin Rush, who practiced in Philadelphia in the early 1800s, is acknowledged as the father of psychiatry in America. Dr. Rush, like Hippocrates, was convinced that mental illness came from a disturbance in the body's four humors, and he prescribed bleeding and purging with laxatives, as treatment for the mentally ill. He was one of the first of his day to popularize the theory that mental illness was caused by something that goes wrong inside the body, rather than being caused by something that enters the body from the outside.

Dr. Emil Kraepelin

The first psychiatrist to make significant progress toward a classification of mental illness was the German physician Dr. Emil Kraepelin. He was born in 1856 and became a professor of medicine at the University of Heidelberg. Dr. Kraepelin began his career as a neurologist, a physician who specializes in diseases of the nervous system, but over time his interest shifted to psychiatry. He started keeping note cards on his patients, documenting their various symptoms, and the course of their disease; and after years of observations he was able to conclude the course that a patient might take with particular symptoms. This is similar to the model that is used in the rest of medicine when doctors use signs and symptoms to make a diagnosis and predict the course of the patient's disease. His note cards eventually became a text book which contained one of the first classifications of mental illness. His textbook also helped revolutionize psychiatry, because for the first time a psychiatrist could predict what the outcome might be for a patient with particular symptoms. The DSM, the diagnostic manual of mental illnesses, started out as an extension of Kraepelin's ideas, but over time its focus was influenced more and more by the tenets of psychoanalysis.

Kraepelin divided psychiatric illness into 13 groups, which included not only the severe mental illnesses, like schizophrenia and manic depressive disease, but also the less severe forms of mental illness, like the neuroses and mental retardation. It was in his classification of the psychosis that he made a series of observations that resulted in a major advance in the diagnosis and treatment of two major psychiatric illnesses. There were two major groups of patients with psychosis, one with fever and one without fever, (he didn't know at the time that the cause of psychosis with fever was syphilis). In the group of patients without fever his observations led to a major breakthrough. He noted that if a patient had severe mood swings, which cycled from highs to lows, the patient would continue to have the mood swings, but would not deteriorate over time. He called this condition manic-depressive psychosis, which is now called bipolar disease. If the patient on the other hand had psychosis without the mood swings, the prognosis was worse, and the patient usually had a progressively downhill course. This second type of psychosis often came on early in life, and was called *dementia praecox*, because it was often associated with the early onset of dementia in young people. One of Kraepelin's students came up with another name for this condition. He called it schizophrenia; from *schizo,* meaning split, and *phren,* which means mind, and this is the term that is most often used today for this illness

Some Thoughts on Schizophrenia

In my experience as a family doctor schizophrenia in the general population is relatively rare. I have a few patients with schizophrenia and they appear to have a condition that reminds me of someone dreaming while they are awake. Untreated they literally cannot tell the difference between their dreams and reality. When their illness is severe, their dream world is their reality, and unfortunately all too often that reality is a nightmare. Patients often describe auditory or visual hallucinations as they are drifting off to sleep, which can be very frightening to them, because they can't be sure whether they are dreaming or are awake. Unfortunately if untreated their hallucinations and delusions often intensify, and become more frequent.

Apart from the difficulty of being able to tell the difference between dreaming and being awake, patients with schizophrenia can have the same types of emotional illness as patients without schizophrenia. They can have anxiety problems which will lead them to have paranoid thoughts, and the feeling that somebody is out to get them. They can have excessive anger which can lead to behavior that can be very dangerous both to themselves and those around them. They can have depression and/or bipolar disease, (the latter is called schizoaffective disorder).

Schizophrenia is not an either/or type of illness and can occur in different degrees and severity. Some patients only have mild hallucinations, or hear faint voices in the background, while others feel like they are trapped in some alternative universe, like the movie *the Matrix*, with no way to escape.

The etiology of schizophrenia appears to have something to do with the neurochemistry of sleep. We know that in schizophrenia there are abnormalities in the brain's chemistry, including elevated levels of a neurotransmitter called dopamine, and abnormalities of another of the brain's neurotransmitters called glutamate. Most of the drugs that are used to treat schizophrenia are aimed at reducing dopamine levels. These drugs are called antipsychotics, or neuroleptics, and they are generally very effective. The real tragedy of schizophrenia is that there are tens of thousands of individuals with this illness who are not getting treated, or who refuse to take medication, and who go through life living in a permanent nightmare.

The symptoms of schizophrenia are sometimes seen in non-schizophrenic states. Parkinsonism, for example, is a disease in which dopamine levels are low, and treatment is aimed at increasing dopamine levels; but sometimes in the treatment of Parkinsonism the medications we use can cause an excessive buildup of dopamine in the brain and this will cause a schizophrenic-like state, including hallucinations and delusions.

Another schizophrenic-like state that also has a relationship to sleep is caused by lack of REM sleep, or dream sleep. It is not uncommon for solo sailors, who sometimes go for long periods without sleep to have hallucinations; and there are numerous reports of sleepy solo sailors seeing islands with palm trees and peaceful lagoons in the middle of the ocean, which will lull them into a peaceful sleep, but when they awake hours later, they realize that they are thousands of miles from land.

Currently, I have three patients in my practice who have schizophrenia. One has only mild symptoms and takes a very low dose of an antipsychotic in order to suppress voices that he sometimes hears in the background. He has never been hospitalized and goes about his daily life unimpeded by his illness. Another has more severe symptoms and has been hospitalized several times. Her schizophrenia is combined with bipolar disease, and on medication she also does very well. She works with her church and is learning Spanish, so she can help volunteer in the local Hispanic community. As long as she stays on medication she remains free of hallucinations and delusions, but several times since I started treating her, she has stopped her medications and within a short time her hallucinations and delusions quickly returned. The third patient has the most severe symptoms and is very paranoid. She hears voices that constantly threaten her and her family. They punish her by causing pain in her back and neck. She has been in the state mental hospital numerous times, and for a time was calling the F.B.I. and the Governor's office on a daily basis. The greatest challenge in treating her was keeping her on medication. Each time after she was released from the hospital, she would immediately stop taking her antipsychotics. She said that the medicines made her feel "funny," and as soon as she stopped her medication the hallucinations would return. When she first came to see me it was not for treatment of her schizophrenia, but it was for the treatment of her back and neck pain, but it was apparent at that first visit that she had all of the symptoms of schizophrenia. I encouraged her to take antipsychotics by mouth, but she refused. I finally said that I would agree to treat her back and neck pain, but only on the condition that she agree to an antipsychotic shot that can be given one a month. She reluctantly agreed, and after she began getting her monthly injections, her delusions and hallucinations markedly decreased. She has remained out of the hospital and she has gotten a job. She is very intelligent, as many schizophrenics are, because schizophrenia has nothing to do with intelligence, a point that was recently illustrated in a movie about a mathematician with schizophrenia.

With each of these patients I have been open about their diagnosis. I explain to them that they have schizophrenia, though most are reluctant to accept the diagnosis. When they come to the office I discuss with them their hallucinations and delusions the same way I talk to a diabetic about their blood sugar levels. Patients with schizophrenia and their families need to know the importance of medication for this illness, and what it takes to keep the patient in this world and out of a world of dreams and nightmares.

Dr. Kraepelin's Goal

One of Dr. Kraepelin's goals was to try and move mental illness onto the same plane as medical illness. He proposed that there were distinct mental illnesses, with courses that could be predicted, like diabetes or thyroid disease, but the one thing that he lacked was an etiology for mental illness. He didn't have an infectious agent, or a chemical abnormality that caused mental illness, and even today we still lack an etiology for most mental illnesses.

Dr. Sigmund Freud

In the late 19th century a split occurred in the psychiatric world between those who believed in the Kraepelinian, or medical model, and those who believed in the psychological model. This split which had been building for some time became more pronounced when a young Jewish physician in Vienna by the name of Sigmund Freud came on the scene. Dr. Freud had started his career as a neurologist, but he turned to psychiatry when he couldn't find an adequate explanation for his patient's symptoms using the theories of mental illness that were popular at the time. Many of his patients were young women from well-to-do Jewish families in Vienna, and he noticed a recurring theme in their histories. There seemed to be sexual conflicts of one kind or another in many of his patients, and Freud speculated that their symptoms must be related to repressed sexual desires, and/or conflicts over sex. He went on to theorize that the patient's repressed desires, or conflicts, must be hidden somewhere outside of the conscious mind, because his patients didn't seem to be aware of them. He theorized that they must lie somewhere in the unconscious, and though Freud was not the first to propose the concept of the unconscious, his emphasis on it gave it new significance.

Over time he went on to develop a road map of the psyche that for years has been accepted as gospel, not only in psychiatric circles, but in the minds of laypersons as well. Freud held that the psyche is divided into three parts. There is a part that is hidden somewhere outside of consciousness which is very powerful and has a great deal of influence on behavior. He called this the "id," which literally means the "it." He felt that the traumas of childhood and sexual repression lay hidden in the id, and when the child becomes an adult these repressed conflicts could arise out of the id and affect behavior without the patient being aware of where the drive was coming from. Freud's id therefore was a very powerful and frightening force. Another part of the psyche was the superego, which he thought was the voice of our parents in our heads telling us what to do, or not to do, and making us feel guilty when we are not perfect. In the middle of these two forces was the ego, or I. The ego was like the captain of a ship trying to steer his or her vessel on a course that he or she had not chosen, with navigational instruments that worked only part of the time, in seas where there was fog and high winds, and where the ship's wheel was only loosely connected to the rudder. It is apparent that in Freud's psychology there is a very strong element of helplessness.

Some of Dr. Freud's contemporaries, especially some of the philosophers, proposed different views of the human psyche. The crossover between philosophical ideas and psychiatric ideas was a common phenomenon in the latter part of the 19th and early 20th century. In fact some of Freud's own writings were as much philosophical as they were psychiatric. The group of philosophers who were strongly opposed to Freud's ideas were the Existentialist, and chief among them was Jean-Paul Sartre. He vehemently disagreed with Freud over the strength of the id and the power of the unconscious. Sartre felt that the ego, or I, must ultimately be given the responsibility for behavior and hiding behind the id is only an excuse.

Through the precepts of existentialism had originated with Soren Kierkegaard in the middle of the 19th century, the movement became very strong in France after World War II. The resurrection of this philosophy came about when a group of philosophers sought to understand why so many of the French had capitulated with the Germans during the war. When the war was over, the capitulators offered various excuses for their behavior, and Sartre and others came to the conclusion that what matters most about an individual is what they do, their behavior, their existence, as opposed to their essence, which is the label that someone has. He felt that all too often an individual, once they have a label, uses the label as an excuse for their behavior. He went on to say that we, as individuals, must be held responsible for our behavior, and we can't blame our behavior on the id, the devil, or a label. There are no excuses.

Psychoanalysis

Over time Dr. Freud developed a type of therapy to deal with the id and the superego that he called psychoanalysis. The aim of psychoanalysis was to sneak up on the patient's unconscious forces and shine a light on what was hidden there in order to expose to the patient the unseen forces that were influencing his or her behavior. During psychoanalysis Dr. Freud would have the patient lie on a couch, while he sat behind the patient out of sight. During the session the patient would talk about whatever came into his or her mind, a process he called free association, which he felt was the best way to enter the patient's id. As the patient talked Dr. Freud would periodically offer an interpretation of the patient's musings, but he soon realized that it wasn't what the patient said that was the most important element of the therapeutic process, it was what took place between the therapist and the patient. He found that over the course of therapy, there would develop between the therapist and the patient a reaction he called the transference. As the transference reaction occurred, the patient would transfer to the therapist the attitudes and feelings that lay hidden in his or her unconscious. Once out in the open the hidden feelings and attitudes could be looked at objectively and interpreted by the therapist. Dr. Freud felt that the insight gained through this process would cause healing and change. In many respects it was like a surgeon opening up a boil to let the pus out and allow healing.

The therapist had to be careful of a reverse reaction call the counter-transference, a condition in which the therapist, if he or she were not careful, might let their own prejudices and feelings contaminate the therapeutic process. As a result of this danger, and the importance of understanding themselves, it was essential that the analyst undergo analysis as well. Dr. Freud was acutely aware of this and actually spent years analyzing himself.

Psychoanalysis was a long and arduous process. It often went on for years, primarily because an end point was never clearly defined. Many patients with "neuroses" were helped with psychoanalysis, but insurance companies eventually became wary of the expense involved and began denying treatment by this method. Unfortunately this prejudice toward psychoanalysis in particular, and the treatment of mental illness in general, is still very prevalent today. This is true despite the fact that with the use of modern drugs and the new psychotherapies, the treatment of most mental illness is often no more complicated or prolonged than the treatment of a medical illness like diabetes, or elevated cholesterol.

Mental illness in Freud's day was divided into major and minor types. The minor illnesses were called neuroses, which literally means disease of the nerves, and is derived from *neuro*, which means nerve, and *osis,* which means abnormal condition. It was generally felt at the time that people with neuroses had a disease of their nervous system and their neurosis was due to abnormalities of the nerves. Psychosis was the term given to patients with more severe mental illness, like schizophrenia and severe manic-depressive disease. The word psychosis comes from *psyche*, meaning the mind or spirit, and *osis*, meaning abnormal condition. The term neurosis is rarely used today, but the term psychosis is still in use.

Eventually it was on neurosis that Freud concentrated, because he discovered early on that psychoanalysis didn't work on patients with psychotic disorders like schizophrenia and manic depressive disease. Within a few years the psychoanalytic movement had become the dominant force in psychiatry; but in their enthusiasm for their discipline, many psychoanalysts forgot the lessons that Freud had learned and tried to fit every psychiatric illness into the psychoanalytic mold, often with disastrous results. When psychoanalytic therapy failed, patients and their families were told that it was because of the patient's resistance to treatment, or it was the family's fault. As an example of their zeal to try and push a square peg into a round hole, the mothers of patients with schizophrenia for years were labeled as being "schizophrenogenic." These mothers were told that it was their fault the child had developed schizophrenia, because they had neglected the patient during childhood. These mothers carried the guilt placed upon them by psychoanalysts for years until, it was shown that the mother's behavior during childhood had nothing to do with the development of schizophrenia, and this disease is caused by abnormalities in some of the brain's neurotransmitters.

The Decline of Psychoanalysis

Even after the antidepressants came along in the 1950s, some diehard psychoanalysts were still trying to treat major depression with talking therapy. During the 1980s there was a famous malpractice suit brought by a patient against a psychiatric institution. The patient, Rafael Osheroff, who happened to be a doctor, had been denied treatment with antidepressants by an analyst at a well-known psychiatric institution. He was being treated with psychoanalysis, but was not getting any better. It was only after the patient transferred to another hospital and was placed on antidepressants by another physician did his depression begin to lift. He continued to improve and eventually was able to go back to work. The patient was so upset that he brought a malpractice suit against the institution. Though the suit was eventually settled by arbitration out of court, the testimony and publicity from the trial dealt a severe blow to psychoanalysis.

I was first introduced to Freud's model of psychology and the concept of psychoanalysis when I was in medical school in the 1950s. One of the things that used to bother me about Freud's model of the psyche was the question I kept asking over and over again, who out there is normal? I remember at the time observing the people I knew, and everybody seemed to be a little neurotic, especially we medical students. We not only had high levels of anxiety, but some of us were even a little paranoid. When I compared my fellow students to our professors, they didn't come out much better. I didn't know of anyone who wasn't a little neurotic. There didn't seem to be anybody I knew that was normal.

In the 1960s and 1970s the treatment of the minor mental illnesses became very popular. This trend coincided with the development of not only newer and more efficient psychotherapies, but also the availability of a new group of more effective drugs, such as the benzodiazepines and the tricyclic antidepressants. One form of psychotherapy that was very popular in the 1970s was called transactional analysis. It was loosely based on Freud's model of the psyche, but with a modern twist. In transactional analysis the id became the child, the superego became the parent, and the ego became the adult. TA, as it was called, was aimed primarily at helping patients with interpersonal relationships and the patient's emotional problems were of secondary concern. There were two books written about transactional analysis at the time that became very popular. One was called *Games People Play*, by Eric Berne, and the other was *I'm Okay-You're Okay*, by Thomas Harris.

In the 1960s and 1970s there was a shift in the location where most psychotherapy took place. It began moving from the psychiatrist's office into the office of the psychologist. In part this was because many of the new psychotherapies were developed by psychologist, but it was also because the treatment of mental illness was shifting away from the psychotherapies to the new drugs. This division of psychiatric treatment into two distinct realms has intensified over the last few decades. One realm is that of the psychiatrist who prescribes and manages medications, but who doesn't do psychotherapy. The other realm is that of the psychologist who does psychotherapy, but who cannot prescribe medication. Many offices today have psychiatrists who prescribe medications, and psychologists who do the psychotherapy. There are a few psychiatrists who still do both, and there are many family doctors who prescribe medications for depression and anxiety, but who generally don't do psychotherapy.

The development of the medications that can successfully treat a wide variety of mental and emotional illness has by any measure been a resounding success, and today most psychiatric illnesses can be treated as easily as medical illnesses. What is missing today is a better foundation on which to base diagnoses, and a better understanding of what treatment is actually doing. Unlike the rest of medicine, psychiatry for the most part has built its house on a foundation of sand, and unfortunately even the winds of political pressure have been strong enough to shift its foundations.

The DSM Series

After Kraepelin's seminal work in the latter part of the 19th century, for years there were only modest attempts to categorize the various types of mental illnesses. In the 1880s the United States Government needed a way to keep statistics on the number and diagnoses of patients who were in mental hospitals; so the government developed a manual with a classification that included mania, melancholia, monomania, paresis, dementia, dipsomania, and epilepsy. When the next manual was written in 1918, psychiatrists became involved, and in addition to being a classification of mental disorders, the manual became a diagnostic tool for doctors. When this edition was published it was called the *Statistical Manuel for the Use of Institutions for the Insane*, and it was used up until World War II. The next edition was called the *Diagnostic and Statistical Manual: Mental Diseases*, and by this time psychoanalysis had become the dominant force in American psychiatry and the influence of psychoanalysts was very much in evidence in the development of the diagnostic criteria for the new DSM. There was also another force at work which resulted in a significant shift in the emphasis on the types of diagnoses that were included. Most psychiatrists by now were working in offices, rather than in mental institutions, so more of the minor mental disorders were included in the list of diagnoses.

The next edition of the DSM was published in 1968 and was call the DSM II, and for some reason or other, the "Statistical" name has continued with each succeeding edition, though as far as I know the government and psychiatrists have gotten out of the business of counting the mentally ill. With each succeeding edition the number of mental illnesses has continued to climb. Occasionally a diagnosis would be dropped, and what was considered a mental illness in one edition might not be considered a mental illness in the next. For example, the validity of homosexuality as a mental disorder, which had been included as a diagnosis in DSM II, was called into question by the gay community when the DSM III was being complied. Because psychiatry had never been able to define normal sexual behavior, when the debate about homosexuality arose, the weaknesses of the foundations of psychiatric theory became all too apparent and the chickens came home to roost. Most psychoanalysts felt that homosexuality was abnormal and deserved to be listed as an illness, and since most of the members of the committee who were writing the next revision of the DSM were psychoanalyst, they felt that homosexuality should be included as a diagnosis. After a spirited public debate the committee backed down, and when the DSM III was published in 1980, homosexuality was no longer considered an illness. The committee hedged on whether homosexuality was normal or not.

Almost immediately after the publication of the DSM III a revision was begun, and in 1987 a revised version was published. This was followed by DSM IV in 1994, another revision a few years later, DSM IV-R, and work is now underway on DSM V. With each edition, diagnoses have been added and deleted based on a committee's decision, but the basic premise of the manual remains the same, and unfortunately the basic weakness of its methodology also remains. Its validity has been called into question in numerous studies, which I believe goes back to the "normal" problem, and its reliability has also been called into question in many studies, which I believe is chiefly because the DSM has never been able to come up with a cause for any of the mental illnesses. The closest it comes to offering an etiology for any mental illness is from the following statement about neurotransmitters in the case of major depression from the IV-R edition.

Neurotransmitters implicated in the pathophysiology of a Major Depressive Episode include norepinephrine, serotonin, acetylcholine, dopamine, and gamma-aminobutyric acid. Evidence that implicates these neurotransmitters includes measures of their levels in blood, cerebrospinal fluid, or urine and platelet receptor functioning. Other laboratory tests that have demonstrated abnormalities include the dexamethasone suppression test, other neuroendocrine challenges, functional and structural brain imaging, evoked potential and waking EEG.

For the most part the DSM series has been accepted by the mental health community as gospel, but over the years there have been a few voices that have expressed concerns about its validity and reliability. One of those is a book entitled, *The Selling of the DSM, The Rhetoric of Science in Psychiatry*. It is highly critical of the methodology used in the formulation of the DSM series, and the book points out many of its deficiencies. More recently I read an article in one of my family practice journals which pointed out another of the DSM's major weaknesses. Often a patient's will shift from depression to anxiety and back again, sometimes they will have both; but the DSM is not set up to deal with what are called co-morbid conditions, where there is more than one psychiatric diagnosis involved. This presents a real dilemma for the family physician, as well as for others who deal with mental illness.

Drugs and Other Treatments

By the middle of the 20th century the limitations of psychoanalysis were becoming all too evident. Three of the most severe mental illnesses, schizophrenia, major depression, and manic-depressive disease, were totally unresponsive to psychoanalysis. The majority of severely ill patients were warehoused in what had previously been called insane asylums, but were now called mental hospitals. The only drug that had had any success in the treatment of mental illness up to this time had been opium, which had been developed by Paracelsus many centuries before and had been used with some success in the treatment of depression. A paper written by Emil Kraeplin at the turn of the 19th century described a regime using opium to treat depression. The opium treatment consisted of giving increasing doses of opium until the patient got better, then the dose was slowly tapered and finally discontinued. Despite concerns that addiction might occur, studies from the period actually reported a very low incident of addiction. Opioids were used periodically to treat depression as late as the 1950s when the tricyclics were introduced, but today opioids are rarely used, except in the treatment of some very resistant cases of depression, and only then in exceptional cases.

Another drug that was used to treat depression around the turn of the century was cocaine. Freud tried it on some of his patients, and even tried it on himself for his own depression, but it too was eventually abandoned because of its addictive qualities.

The story of how we got from the latter part of the 19th century, when there were almost no drugs, to the present day, when we have a multiplicity of drugs, is a very fascinating story, and one of the better books written on the subject is Dr. Edward Shorter's *A History of Psychiatry*. Dr. Shorter is a professor of medical history at the University of Toronto in Canada. I found his book to be an excellent history of psychiatry, though some psychoanalysts might take issue with his views on the limitations of their discipline.

Up until the early part of the twentieth century, many cases of severe mental illness were caused by the late effects of syphilis on the brain. Syphilis is a very insidious disease with many manifestations. It is caused by a corkscrew shaped organism, called a spirochete, and is a venereal disease. It was often a difficult disease to diagnose, and up until reliable diagnostic test were developed in the last century, the only way to make the diagnosis was at autopsy. When I was in medical school, penicillin, which can easily treat syphilis, had been available for less than 20 years, and we would occasionally see a patient with some manifestation of syphilis, but by then it was becoming rare. We were taught in medical school that there had been a time when a physician who could treat all of the manifestations of syphilis, would know all of medicine.

The Fever Cures

In 1883 an Austrian psychiatrist by the name of Dr. Wagner-Jauregg noticed that one of his psychotic patients went into remission when she got erysipelas. Erysipelas is a severe skin infection which is cause by the streptococcus bacteria and is associated with a high fever. He reasoned that maybe it was the fever that had caused her improvement, so he began looking for ways that he could artificially produce a fever and test whether it was the fever that made his patient better. He knew that a new vaccine, which had recently been developed against tuberculosis had fever as a side effect, so he decided to try the vaccine on some of his psychotic patients and deliberately give them a high fever. To his amazement some of his patients went into complete remission and remained that way for some time. What he didn't realize was that the patients who got better were patients with syphilis, and the ones who didn't get better didn't have syphilis. The reason the patients with syphilis were getting better was because the fever was killing off the spirochetes, which was causing the psychosis.

Unfortunately the tuberculin vaccine sometimes had severe side effects, so Dr. Wagner-Jauregg decided to look for another form of fever therapy with fewer side effects. A well-known cause of fever at the time was malaria, and the standard treatment of malaria was to administer quinine as soon as the patient developed fever, which would immediately bring the fever down. When one of his psychotic patients, a soldier who had picked up malaria in the tropics, started running a high fever, he decided to let him develop the full-blown disease by withholding quinine. After nine febrile episodes he finally gave him quinine, but by then the patient's psychosis had completely cleared and the patient was symptom-free. Dr. Wagner-Jauregg went on to treat other patients with the fever cure, and some of them also had remissions. When the word got around that the fever cures helped some patients with psychosis, this cure became very popular, even though psychiatrists didn't realize at the time the patients who were getting better had a mental illness that was being caused by an infection.

The success, albeit limited, of Dr. Wagner-Jaureeg's experiments had a profound influence on the attitude that psychiatrist had toward the severely mentally ill, because prior to this time nothing that had ever been tried on patients with severe mental illness had ever made any difference. Now for the first time there was hope that something could be done to help some of these unfortunate souls.

Soon syphilis as a cause of madness would be on the wane. In the early part of the 20th century a German doctor, whose name was Paul Ehrlich, discovered an arsenical compound that could treat syphilis. He called it "606," because it was the 606th compound that he had tried before finding one that was effective, and that didn't have severe side effects. It was not long after the discovery of "606" that an even more momentous discovery was made by Sir Alexander Fleming. In 1929 he noticed a curious phenomenon in his bacteriology lab that he had never noticed before. On one of his culture plates there was a ring where bacteria should have been growing, but inside the ring, it was clear. In the center of the ring he notice a mold growing that had somehow gotten onto the culture plate. Rather than throwing it in the trash, he decided to try and figure out why there were no bacteria growing in the ring around the mold. He found that the mold producing this unusual phenomenon was from the penicillium family, and he and others eventually isolated the antibiotic penicillin from this mold. His chance observation, and subsequent curiosity, would usher in the modern era of antibiotics, and infections such as syphilis and other bacterial infections, which had killed millions, would soon be brought under control. Though syphilis is occasionally seen today, it is a relatively minor cause of mental illness. Today there is another infection that can cause mental illness that is even bigger than syphilis was in its heyday. AIDS, which is caused by HIV, can cause hallucinations and delusions in its late stages.

Although there were a number of drugs that had been tried in the treatment of mental illness in the early part of the 20th century, none had met with any appreciable success. Opium and cocaine, as mentioned earlier, were used in the latter part of the 19th century, and the early part of 20th century, but their use was limited because of side effects. Another popular treatment at the time was the use of laxatives. Cathartics had been used for centuries because it was felt that mental illness was caused by some kind of toxin, and the hope was that laxatives could flush the toxin from the body. As one might imagine, laxative treatment for mental illness was not a great success.

The Early Drugs

The first real success of a psychotropic drug which had any effect on mental illness came in the early part of the 20th century. Chemists in Germany were experimenting with a group of drugs derived from plants called alkaloids. The first alkaloid that had any effect on mental illness was the drug hyosycamine, which comes from the belladonna plant. When it was tried on mental patients, it seemed to help a little, but its effects were minimal at best.

Another compound that was synthesized about this same time was a sedative called paraldehyde. It helped calm a patient's anxiety, but it could be addicting, and one of its side effects was a peculiar odor. If someone was taking it, everybody around them knew it. I remember using paraldehyde as an intern in the emergency room, where we used it to treat DTs, but I haven't heard of anyone using it in a long time.

In the early part of the 20th century, a London physician wrote a paper documenting the successful treatment of a patient with epilepsy with one of the fundamental elements, bromide. Some of the asylum doctors decided to try bromide on their patients, because of its apparent calming effect, and it did help to calm patients somewhat. Dr. Neil Macleod, another English physician, decided to carry this treatment one step farther when he discovered that if he gave his patients enough bromide he could put them to sleep. He noted that after a period of sleep some of his patients seemed to feel better. He decided that if a little sleep was good, maybe more sleep was better, so he decided to try putting some patient to sleep for several days and see what would happen. Some of his patients with psychosis did get better, and a few seemed to clear completely. Sleep therapy also worked fairly well for mania, and to a lesser degree depression. Unfortunately sleep therapy was not without its risk, because it was hard to judge exactly what the therapeutic dose should be, and if the dose was too high, there was a risk of over-sedation, brain damage, pneumonia, and death.

After the success with bromide, other elemental salts were tried to see what effect they might have on mental illnesses. One of those was lithium, which is close to bromide in the periodic table. Lithium proved to be even more effective in the treatment of mania than bromides. It had fewer side effects and is still widely used today in the treatment of bipolar disease.

In the early part of the 20th century in Europe, and especially in Germany, chemist were synthesizing a wide variety of chemical compounds to see what effect they might have on mental illness. As soon as they came up with a new one, they would give it to psychiatrists to try on their patients. In one of those experiments, two German chemists took a compound that had been synthesized some 50 years earlier by another chemist, and modified it to see what effect it might have on patients. The original compound had no sedative effect on patients, but when the modified compound was tried on patients it was found to have a strong sedative effect. The earlier chemist had named his compound after one of his girlfriends, whose name was Barbara, and this new class of psychotropic drugs was given the name barbiturates. The barbiturates were found to have a distinct advantage over the other sedatives in use at the time, and soon both short-acting and long-acting forms were developed.

When the barbiturates came into use during the 1930s, a new interest in sleep therapy developed when the barbiturates turned out to be safer than bromides. Barbiturates are still in use today, primarily in the treatment of epilepsy, and as rapid acting sedatives, but their use has declined in the treatment of "nerves" because of their tendency to cause dependence, and because the newer drugs, like the benzodiazepines, had fewer side effects, and were generally more effective.

The Coma and Shock Therapies

Soon a new form of therapy came into vogue which involved even deeper sleep. It was called coma therapy. It was first used by a young Austrian physician by the name of Dr. Manfred Sakel. Dr Sakel's patients were primarily addicts, but in addition to those patients, he also had patients with more severe psychiatric illnesses. One day he noticed a peculiar thing about one of his diabetic patients who was addicted to morphine. The patient got his regular shot of insulin, but didn't eat. His blood sugar dropped and the patient went into insulin shock. When the patient recovered Dr. Sakel noticed that his craving for morphine was greatly reduced. So he decided to see what would happen if he gave one of his addiction patients without diabetes a dose of insulin and deliberately cause their blood sugar to drop. The patient went into insulin coma as predicted, and when the patient came out of the coma, he too seemed better.

Dr. Sakel then decided to try insulin shock on some of his non-diabetic patients who were psychotic. The result on some of these patients was even more dramatic. When some of them woke from the coma they were like different people. Some of the patients with depression improved, and some of the patients with schizophrenia also got better. It was the first time that any treatment for schizophrenia had shown any success.

This treatment sounded almost too good to be true, and not surprisingly there was considerable skepticism in the psychoanalytic community. Disciples of Dr. Freud had a hard time reconciling coma treatment with their ideas of the id, ego, and superego, but despite the resistance of the psychoanalytic school, insulin shock therapy quickly caught on and was soon being widely used. It was the only thing at the time that helped severely ill patients, and when I was a psychiatric resident in the late fifties, we were still using it on select patients.

But insulin shock therapy was not without danger; it was difficult to predict just how low the blood sugar would drop, and if the blood sugar dropped too low, and for too long, permanent brain damage could result. Occasionally convulsions would result, and though at the time convulsions were not felt to be desirable, as we shall see, the aim of the next treatment regime was to actually induce convulsions.

In Budapest, a physician by the name of Dr. Meduna was doing research on patients with schizophrenia and epilepsy. It had been noted by him and others that patients with epilepsy rarely developed schizophrenia, and patients with schizophrenia seemed to have less epilepsy. He wondered if a patient with schizophrenia would get better if they had seizures. He decided to try out his theory on a man with delusions. The man heard voices in his stomach and had quit eating. Dr. Meduna decided to treat him with a drug called camphor, which has seizures as a side effect. After the patient had five seizures, he awoke as if from a dream and began talking normally and asked for something to eat. The year was 1934.

The search for other drugs that would cause seizures eventually lead to the use of Metrazol, another seizure-producing drug, and for a few years it was the primary drug used to produce seizures until a new form of seizure therapy came into use which did not include the use of drugs. In Rome a physician, whose name was Dr. Cerletti, was carrying out research on epilepsy with dogs. He was experimenting with electric shock as a way of finding out more about the cause of epilepsy. His first experiments failed and the dogs died, but he found that if the electrodes were placed on the dog's temples, he could produce a seizure, and the experimental animal would recover with no apparent side effects. He wondered if producing a seizure by electric shock in a patient with schizophrenia would help them. With great reluctance he and some of his colleagues decided to try it on their first patient in 1938. The patient was an engineer with hallucinations who had been found wandering around the railroad station and was refusing to eat. After five treatments the man was well enough to go home. Some of his delusional voices returned, but he was still able to function. It wasn't the perfect cure for schizophrenia, but it was a major advance. Dr. Cerletti called his new treatment "electro-shock," and this new form of therapy soon crossed the Atlantic where it was greeted with great enthusiasm.

It was soon discovered that the real benefit of electro-convulsive therapy was not in the treatment of schizophrenia, but instead was in the treatment of major depression. For major depression it was like a miracle. Up until then, except for some varied success with opium and the other shock therapies, there had never been any treatment for major depression that proved to be as successful as ECT. It also proved to be effective in patients who had depression coupled with mania, and became safer when muscle relaxants were used in conjunction with it. When I was in training at State Hospital in 1958 it was the treatment of choice for major depression. ECT is still used widely today in selective cases of treatment-resistant depression. It is a very safe and effective way to treat major depression when drug therapy fails.

Though ECT was generally accepted by the public, there was one form of treatment during this era that probably contributed more to the apprehension, and negative feelings about psychiatry than any other form of treatment before or after. It was called prefrontal lobotomy. In this procedure a device that resembled an ice pick was driven into the bottom of the brain through the orbits above the eyeballs and swept from side to side, severing the nerve tracts that go from the front of the brain to the back. The patients became calmer after treatment, but unfortunately their spirit and individuality was lost forever. I remember seeing patients on the wards at State Hospital who had had prefrontal lobotomies, and they seemed more like zombies than human beings.

The Discovery of the Neurotransmitters

The first real breakthrough in the treatment of psychiatric illness occurred in the 1950s with the discovery of a series of drugs that affected the neurotransmitters. But, first the story of the neurotransmitters. In the 1920s scientist from around the world were trying to figure out how nerves were able to transmit signals from one part of the body to another. In Germany a physiologist by the name of Herman von Helmholtz had made the observation that when a nerve is cooled transmission of the impulse slows down. He reasoned that since a chemical reaction also slows with a drop in temperature; perhaps some type of chemical reaction was involved in nerve transmission. He himself was not able to solve the problem, but another German physiologist was able to devise an experiment that proved that there is a chemical reaction involved in the transmission of messages by the nerves.

In his book *Mysteries of the Mind*, Dr. Richard Restak, a neuro-psychiatrist, tells the fascinating story of how a German physiologist, Otto Loewi, was finally able to show that nerve transmission does indeed involve a chemical reaction. Loewi had been working on this problem for some time, but hadn't been unable to come up with an experiment that would prove his hypothesis. One night an idea for an experiment came to him in a dream. He woke up and wrote something down, but the next morning he couldn't read what he had written down. When he had the same dream the next night, this time he became wide awake and wrote down the experiment in detail.

In his experiment he took the beating hearts from two live frogs, with the nerves attached, and placed them in a saline solution. He then applied an electrical stimulus to the nerve of one of the hearts and it slowed down, but the other heart also slowed. Loewi concluded that the first heart must have released a chemical that flowed through the saline solution and caused the other heart to slow also. The chemical was later shown to be acetyl choline and was the first neurotransmitter to be discovered.

Neuroanatomy had shown that the nerve cell, or neuron, has a central nucleus with numerous fine threads extending out from it that connect to other neurons. The one that carries the nerve signal out from the neuron is called the axon, and it can be up to several feet long. The transmission of the impulse down the axon is electrical, not chemical, and is caused by an electro-chemical reaction that causes electrons to flow down the axon. The axon has a white fatty layer of insulation around it which prevents the electrical impulse from being short-circuited. In multiple sclerosis a yet unknown disease process causes the insulation surrounding the axon to break down, which results in short-circuiting of the nerve impulse and can cause weakness and paralysis. Extending out from the neuron like the roots of a tree are other threads called dendrites. The dendrites are like antennae that receive the incoming signals from the axons of other neurons. Where the axon meets the dendrite there is a synapse or junction, and it is at this junction where the nerve impulse changes from electrical to chemical. The chemical compounds that carry the signal across the junction and change the signal from electrical to chemical, are called neurotransmitters. When the electrical impulse reaches the end of the axon at the synaptic junction, a burst of neurotransmitters is released from the pre-synaptic side of the axon. The neurotransmitters cross over to the post-synaptic side of the dendrite and fit into receptor sites much like a key fits into a lock. When this happens a new electrical charge is created and the message becomes electrical again. The message then moves up the dendrite, through the neuron's nucleus, and out the other side, down that neuron's axon to the next dendrite, and so on. After the new electrical signal is created, the neurotransmitter is released from the receptor site on the post-synapse side, moves back across the synapse to the pre-synaptic side, where it is stored until another impulse comes down the axon and stimulates it to be released again.

After the neurotransmitters were discovered, it didn't take long for scientist to figure out that this process could be manipulated. They found that certain drugs could jam, or partially jam the receptor site, thus blocking the message from going through, or decreasing its intensity. Blocking drugs were called antagonist. Scientist also discovered other drugs that could mimic a particular neurotransmitter and caused the same or greater effect. These drugs were called agonists. There were also some drugs that could act both as an agonist and as an antagonist. In small doses they would act as an agonist, but in higher doses they would jam the receptor sites and act as an antagonist.

The first neurotransmitter to be identified was acetyl choline, the neurotransmitter that made the heart slow in Otto Loewi's experiment, and it was discovered that the receptor site for acetyl choline could be blocked by a tincture made from the belladonna plant. The word belladonna in Italian means beautiful girl, and in ancient Rome eligible ladies knew that if the leaves from this flower were dissolved in a liquid and drunk, it would make the pupils of the eyes dilate. They intuitively sensed that dilated pupils are perceived to be more attractive to the male than narrowed pupils, and recent psychological studies have shown this to be true.

The effects of belladonna were found to be caused by a chemical called hyosycamine. Hyosycamine will fit into the acetyl choline receptor site, block it, and act as an antagonist. Acetyl choline is the neurotransmitter that is found primarily in the parasympathetic nervous system, the system which controls many of the body's internal functions, such as the intestines and the pupils of the eyes. Though it is not used very often today, I still use tincture of belladonna occasionally in my practice. It is an excellent antispasmodic and I have found it to be very effective in the treatment of irritable bowel syndrome. It is necessary to adjust the dose because an excessive amount will cause the patient to have a dry mouth and their pupils to dilate, but once the patient finds the right number of drops it works well.

The First Effective Psychotropic Drugs

Most of the discoveries in psychiatric drugs have been in the field of neurotransmitters, and the work on the first psychotropic drugs was usually a hit or miss proposition, and often a study in serendipity. The story of the first drug that truly revolutionized psychiatry goes back to 1937, when one of the scientists working for the pharmaceutical company Rhone-Poulenc in Switzerland discovered a new class of drugs that blocked allergic reactions. Since histamines were known to be a part of the allergic reaction, this group of drugs was called antihistamines, and it turned out that one of the side effects of the antihistamines was sedation. Benedryl a very popular antihistamine, has sedation as a side effect and is added to Tylenol and marketed as Tylenol PM as a sleep aid.

When the original antihistamines were tried on psychiatric patients, the results were disappointing. The scientist at Rhone-Poulenc kept altering the antihistamines to see if their psychotropic properties could be improved, and in 1951 they came up with a new class of antihistamines called phenathiazines. One of the effects that they were looking for in this new class of antihistamines was the potentiation of anesthesia without excessive sedation. Paul Charpentier, Rhone-Poulenc's chief phenathiazine expert, gave a sample of one of the new phenathiazines to a French navy surgeon, by the name of Dr. Henri Laborit, and suggested he try it on some of his surgical patients to see if it had less sedation than some of the other drugs. When Dr. Laborit tried it on some of his patients he was struck by something that he had not seen before with the other antihistamines, it produced an unusual calming effect. The patients had what he called in French, a certain *"disinteressment,"* and this effect occurred without the sedation usually associated with previous antihistamines.

One day over lunch in the hospital cafeteria, he suggested to three of his psychiatric colleagues that they try this new antihistamine on some of their psychiatric patients. The drug was called chlorpromazine and the first patient that it was tried on was in the manic phase of bipolar disease, and he seemed to get a little better. He was also given phenobarb and electric shock, and was soon able to function on the wards. He began eating and bathing himself, which he had not been able to do before. Though he was still somewhat hypomania he was clearly better with this treatment than with anything else that had been tried. The word soon got around in the psychiatric community in Paris that there was a new drug that looked promising. While its effect on mania proved to be an improvement over previous drugs, when it was tried on patients with schizophrenia, its true worth was discovered. The effect that it had on patients with schizophrenia was like a miracle. Many of the patients who had been locked away on the backwards for years, delusional and unable to function, awoke as if from a nightmare and returned to the real world. The first large study with chlorpromazine was done at Ste. Anne Mental Hospital in Paris on a group of patients with schizophrenia in March of 1952. The effect was so dramatic that the study was published a scant three months later in June of that year and the use of chlorpromazine quickly spread throughout the French mental hospitals. A newspaper article in Paris proclaimed in May of 1953,

....the atmosphere in the disturbed wards of mental hospitals in Paris was transformed: straitjackets, psycho-hydraulic packs, and noise were things of the past! Once more, Paris psychiatrists, who long ago unchained the chained, became pioneers in liberating their patients, this time from inner torments, and with a drug, chlorpromazine. It accomplished the pharmacologic revolution of psychiatry.

It would be a number of years before the real impact of this revolutionary accomplishment would become apparent, but the reality was, if a chemical such as chlorpromazine could be used to treat schizophrenia, then the etiology of schizophrenia must be chemical. Up to this point Freud's theories about psychiatry and psychotherapy had been held as gospel, but when this new drug came into use there were no theories forthcoming about how this drug could affect the id, ego, and superego.

Since the time of Descartes, who actually got it from Plato, there had always been the assumption that the mind and body were two separate entities, the so-called mind-body dualism; but if a chemical could affect the mind in so dramatic a fashion, then the mind itself must have a chemical foundation. Today the mind is considered by most scientists to be a product of the brain, and when the brain is working properly, it will produce a good mind; but when it is not working properly, the mind will be defective.

To date there is no theory that explains the relationship between the neuro-chemical aspects of the brain, and the psychological and emotional aspects of the brain. Such a theory would explain how both psychotherapy and drugs can both work in the same illness. We know, for example, that psychotherapy and antidepressants can both treat some types depression, but we still don't understand the relationship between the two, and whether depression is a psychological problem, or a chemical problem, or both. One of the projects of this book is to present a paradigm that offers an explanation for this puzzle.

The new antipsychotic drugs soon crossed the Atlantic and when I was working at State Hospital in 1958, doctors were just started to use the new antipsychotic drugs. The first of these drugs was chlorpromazine, the drug used in the Paris studies, and whose brand name was Thorazine. It was soon followed by other phenathiazines, notably Stellazine and Compazine and this class of drugs was given the name neuroleptics. Subsequent research showed that the neuroleptics worked by reducing dopamine levels in the brain. Unfortunately serious side effects began to show up in some patients. They began to walk with a peculiar gait, and some developed an uncontrollable movement of the tongue. The condition was given the name tardive dyskenesia, and it became a major deterrent to the use of the new neuroleptics. Most of the time the patient would get better when they were taken off these drugs, but unfortunately in a few patients the symptoms became permanent.

The other major shortcoming of the neuroleptics was a lack of effectiveness in the number one major mental illness, depression. While ECT was usually successful in treating major depression, there were no drugs available to treat the less severe forms of depression, and although schizophrenia is a devastating illness, it is relatively rare, while depression, both major and minor forms, is very common. Depression takes an enormous toll on society in terms of the misery that it causes, and from the deaths in patients who commit suicide.

After the discovery of the phenathiazines, chemist in the labs continued to experiment with the phenothiazine molecule, hoping to come up with new drugs that might be an improvement over the ones that existed. When two atoms in the phenothiazine molecule were changed, they came up with a compound that was called imiprimine.

When this new compound was tried on patients with schizophrenia, it didn't help; in fact it sometimes made them worse, but when it was tried on patients with depression, the results were a different story. Slowly the patient's depression began to lift and within two to three weeks many patients were free of a burden they had been carrying all of their lives. The miracle that had occurred with chlorpromazine with schizophrenia was repeated with imiprimine in depression. Imiprimine was given the trade name Tofranil, and for the first time in the history of psychiatry, except for opium, there was a drug that could effectively treat depression. When it was tried on patients with anxiety, it was also shown to be an effective anti-anxiety drug.

Soon there would be other drugs in this class, just as there had been with chlorpromazine. This new group of antidepressant drugs was called tri-cyclics, because the basic molecule had a three-ring structure. The second one to be developed was called amitriptyline, whose trade name became Elavil. Both it and imiprimine are still widely used today. The new antidepressants brought about a revolution in the treatment of depression, which up to this point had resisted treatment; but unfortunately this new class of antidepressants, like the first antipsychotics, also had side effects. They could cause dry mouth and constipation, and an overdose could be fatal. It would be a number of years before safer antidepressants would be developed, but in the meantime, many patients with depression could be treated with medication, even though the first antidepressants carried the risk of serious side effects.

The First Specific Anti-anxiety Drugs

As the drug treatments became more successful in the 1960s, the attitude of society toward mental illness slowly began to change, and some of the stigma that had for so long been attached to it began to melt away, though it has by no means completely disappeared some 40 years later.

As the profits rose through the sale of the new psychotherapeutic drugs, the pharmaceutical industry saw a potential market in the treatment of the one mental illness that lacked an effective drug, anxiety, and they turned their attention turned toward a problem that was even more common than depression. The only drug class available at the time was the barbiturates. Barbiturates had been around since the turn of the century and were regularly used to treat cases of "nerves," as anxiety was often called. I used them regularly in the 1960s, because at the time it was all we had, but phenobarb and the other barbiturates had a number of undesirable side effects, such as sedation and dependency, and an overdose could be fatal. The drug companies therefore had an incentive to come up with better drugs to treat this very common problem.

In the late 1950s a new drug that could treat anxiety was developed called meprobamate. It was marketed under the trade names Miltown and Equanil. Meprobamate was very similar to phenobarb, and it too could be overly sedating, and patients often became dependent on it. It passed into obscurity within a few years. After meprobamate, the pharmaceutical companies found themselves at a dead end in their search for new drugs to treat anxiety, but just around the corner there was a major breakthrough with the discovery of a new class of anti-anxiety drugs called the benzodiazepines. How the benzodiazepines were discovered is truly a story in serendipity.

In the late 1930s Roche Pharmaceutical had a Jewish scientist from Poland named Leo Sternbach working for them in Switzerland. When World War II broke out Roche sent him to the United States, because they were afraid that the Nazis might invade Switzerland and take him back to Germany. In Switzerland he had been working on a project to find a drug that might have some effect on anxiety, and when he got to the States he was allowed to continue the same project. He was working with a group of chemicals called alanines from which dyes are formed. As far as I know he had chosen this chemical group at random. He was just fishing, a technique often used in research. He would synthesize a compound from this class and give it to the researchers in the animal lab to see what effect this new compound might have on test monkeys. After many years of research he had turned up nothing. His supervisors at Roche finally told him to give up on this project and move on to something else. As he was cleaning out his lab, he noticed some crystals in the bottom of one of the test tubes that he had not noticed before. He sent this compound over to the animal lab for one last try.

A few days later he got a call from the lab that his latest compound seemed to show some promise. When it was tried on a group of monkeys which were known to be very aggressive, they immediately calmed down. The researchers noticed something else about this compound that was different from the other anti-anxiety agents, with this new one the monkeys appeared to be calmer, but they were still alert; whereas with other anti-anxiety agents they were usually quite sedated. This new compound was given the generic name chlordiazepoxide. In 1959 the first human trials were held and the patients reported that their anxiety was greatly relieved, and they were able to sleep better. Yet they still felt alert without the sedation usually associated with phenobarb. A larger human trial was carried out, and the second trial confirmed what the initial study had shown. This new drug appeared to be more effective and safer in the treatment of anxiety than anything that had been used before. When it was brought before the Federal Drug Agency in 1960, it became the first drug to be approved specifically as an anti-anxiety agent. Chlordiazepoxide was given the trade name Librium and it quickly became the number one selling drug in the United States. The financial incentive that had driven Roche to look for a drug to treat anxiety began to pay off in a big way.

Librium was a definite improvement over the other drugs that had been used to treat anxiety, but it also had some undesirable side effects, the most noticeable being that it could cause seizures if it was stopped abruptly. Roche asked Sternbach to go back to the lab and see what other compounds he might come up with from this same family. After a short time he came back with a related compound that was given the generic name diazepam. It became even bigger than Librium, and is better known by its trade name, Valium. Valium quickly surpassed its sister drug in sales, and until Prozac was introduced 13 years later; it was the most successful psychotropic drug to ever hit the market.

No one had a clue as to how these drugs, derived from a dye compound, could have such a profound effect on anxiety. Researchers began an intense effort to try and understand how the benzodiazepines worked. They reasoned that there must be some kind of benzodiazepine receptor site in the brain, and after an arduous search they found one. The implications of this finding in the understanding of anxiety was enormous, for if there were benzodiazepine receptor sites in the brain, the brain must either make a benzodiazepine neurotransmitter, or something very similar to it. The discoveries of the benzodiazepines and their receptor sites forced psychologists and psychiatrists to rethink not only their ideas about anxiety, but about the emotions in general. The old adage that an emotion like anxiety was, "all in your head," was rapidly becoming untenable. Just as when the first antipsychotics and the first antidepressants were introduced, the new anti-anxiety agents only added to the evidence that all of our emotions have a chemical basis.

The SSRI's and Beyond

Though a number of psychotropic drugs were introduced after Valium, such as buspirone (Buspar), another anti-anxiety agent, and buproprion (Wellbutrin), another antidepressant, the next major breakthrough came with a new group of drugs, the selective serotonin reuptake inhibitors. The primary purpose of the SSRI's is to increase serotonin in the synapse, and they include drugs like Prozac, Paxil, and Zoloft. Though they were developed primarily as antidepressants, they were also found to have anti-anxiety properties. A major advantage of the SSRIs is they were found to be much safer than the older tricyclics antidepressants. If a patient overdosed on them it was not fatal. Occasionally they caused side effects such as abdominal pain and sexual dysfunction, but their side effects were generally mild, and for the most part the SSRIs soon replaced the older tricyclic antidepressants.

The story of the SSRIs goes back to 1953 when John Gaddum, a Scottish researcher, speculated that one of the brain's neurotransmitters, serotonin, might be involved in mood regulation. At the same time in this country, at the National Institute of Mental Health in Bethesda, researchers were also doing similar studies on serotonin. It had been known for some time that one of the older antihypertensive drugs, reserpine, had an effect on serotonin; and one of the side effects sometimes seen with reserpine was severe depression. The parent drug of reserpine was rauwolfia, which came from a root grown in India. In 1963 Alex Coppen, an English researcher, tried a drug very similar to serotonin on a patient with depression. The patient got some better, but the drug had many side effects, and the direction in research soon turned from using serotonin-like drugs, to drugs that would cause a serotonin effect. In 1981 Arvid Carlsson at Astra Pharmaceuticals in Sweden came up with a drug that did this by inhibiting the reuptake of serotonin in the brain's synapses, and keep serotonin in the synaptic junction for a longer period of time, causing a greater serotonin effect. The first drug to be tried was called Zelmid, but it had too had many side effects and was soon withdrawn from the market. After the unfortunate experience with Zelmid, research on the SSRIs was put on the back burner. At Lilly Pharmaceutical, an SSRI with the generic name fluoxetine had been developed, but the company's leadership had little interest in exploring its use in depression because of the prior experience with Zelmid. Two researchers at Lilly in this country, Ray Fuller and David Wong, who had been following the research in Europe, went to the leadership at Lilly and convinced them to pursue further testing of fluoxetine. It was eventually put into a clinical trial, and the initial results on the treatment of depression looked promising. Larger studies were set up, and these confirmed the results of the initial trials. One of the advantages of fluoxetine was a lack of sedation, and an overdose was not fatal. In 1987 fluoxetine was released under the brand name Prozac, and like Valium, it too became a household name. Soon other versions of the SSRIs were developed and included drugs like Paxil, Zoloft, and Celexa. In addition to their antidepressant qualities, the SSRIs were also found to have anti-anxiety qualities, just like the tricyclics.

Following the development of the SSRIs, research has continued in an effort to find new and better psychotropic drugs. One of the latest groups is called atypical antipsychotics. They were originally developed to treat schizophrenia and the psychosis associated with mania, but since their introduction several years ago, they are now used "off-label" to treat other conditions such as dementia, bipolar disease, and more recently depression. Though they were initially felt to be free of side effects, unfortunately weight gain and diabetes began showing up in some patients, again demonstrating there are very few perfect drugs.

Another group of drugs that have been developed within the past few years affect more than one neurotransmitter in the synapse. Drugs like Effexor and Cymbalta block the reuptake of both serotonin and norepinephrine, and increase the effect of both of these neurotransmitters in the synapse. Both of these drugs have also been shown to be efficacious in the treatment of anxiety and depression.

After the SSRIs were found to be effective in the treatment of depression, initially it was assumed that the cause of depression must be a lack of serotonin, but we now know that any number of drugs can treat depression, not just those that increase serotonin, and it is becoming apparent that the neurochemistry of depression is far more complicated than was once thought.

Though the influence of the SSRIs on the treatment of both anxiety and depression has been enormous, there has been a concern on the part of some lay people, as well as some physicians, that perhaps that the SSRIs were making people feels too good. Dr. Peter Kramer several years ago wrote a book, *Listening to Prozac,* in which he expressed this idea. When a patient is psychotic and hallucinating, it is easy to say when a patient is ill, because delusions and hallucinations are obviously abnormal, and when we treat a patient with those symptoms we know what our goals are. But when it comes to the treatment of problems like chronic anxiety or minor depression, we don't have very good guidelines. What are the therapist's goals for these conditions? Is it to rid the patient of all anxiety? That wouldn't be good, because there are times where we need some anxiety. Is depression ever normal? Are there any criteria that we can use to say when feelings are normal? How are we supposed to feel?

When I couldn't find the answers to these and other questions, I began a search for answers that were more rational than the ones that we have now. It seemed to me that the major voids in the understanding of mental and emotional illness were a definition of normal and a rational cause for mental illness. The rest of this book is about my search for things about the emotions that can be measured, arriving at a definition of normal, and finding a cause for abnormal emotions. But first, with the reader's indulgence, a bit of my own personal story and background

Medical School, Internship and an Unexpected Year in Psychiatry

When I entered Wake Forest College I originally intended to take two years of pre-optometry, transfer to an optometry school, and become an optometrist like my father and grandfather before me. My grandfather had been one of the first optometrists in the state, and my father had taken over his practice when he retired. Sometime in the fall of my junior year I decided that I didn't want to fit glasses the rest of my life, and instead decided to apply to medical school and become a doctor. This was what I had really wanted to do every since I was a boy growing up on a small farm outside of Raleigh, N.C. My overall grades in college were good, but not great. My grades in the sciences were much better than average, and I had been the manager of our championship basketball team, sports editor of our college newspaper, and class president, and I hoped that the admission committee would look on my application favorably. I breathed a sigh of relief when they did, and in the fall of 1953 I began medical school at Bowman Gray School of Medicine in Winston-Salem, N.C.

In medical school we were taught the basics in medicine, surgery, pediatrics, and a smattering of Freud and psychoanalysis. We learned a little about neurosis and psychosis, and we made rounds at the local psychiatric hospital where we saw patients with schizophrenia and other major psychosis. But the treatment of severe mental illness at the time was limited to the shock treatments, insulin and electric. In retrospect psychiatry now seems to have only been a footnote in our medical school curriculum.

I graduated from medical school in the summer of 1957 and left for an internship at Grady Memorial Hospital in Atlanta. The "Grady's," as it was affectionately called, was the large public hospital that served Atlanta and the surrounding counties. The teaching staff came from Emory University, and Grady had the reputation of being one of the finest teaching hospitals in the country. The hospital itself like so many other big city hospitals of that era was old and neglected, but across the street a new hospital was being built to replace it, and we were told that we would be moving into the new hospital in the spring. Those were the days before integration, and the hospital was completely segregated. We had the "white side" and the "colored side," and each side had its own wards and emergency room. The only thing that wasn't segregated was the operating rooms. We interns were furnished room and board, uniforms, laundry, and a monthly stipend of twenty-five dollars per month. That year many senior medical students opted to go to hospitals that paid more, and we wound up with about half the house staff we normally should have had. Those of us who did come had to do double duty that year. The administration took pity on us and tripled our salary to seventy five dollars per month. The next spring we moved into the new hospital as promised, and the patients and staff all thought that we had died and gone to hospital heaven. Everything was brand new. In the old hospital, the interns and residents were two or three to a room, but now we had our own individual rooms, and though the rooms were small, they had new furniture and a nice view of the city of Atlanta. We had a shiny new cafeteria where the hospital staff took their meals, and the operating and emergency rooms had the latest in equipment. It was as though we had gone from the late 19th century to a modern era overnight. But we still had the "white side" and the "colored side." That spring we got hit by the worse flu epidemic since 1918. Many people died, and we often had to work 24 hours straight.

I had signed up for a mixed internship, which meant the year was divided up into internal medicine, obstetrics and gynecology, and the emergency room. My plan was to do a mixed internship, go into the Army for two years, take another year of training in pediatrics, and then go into family practice. I worked hard at Grady that year and I learned more than most interns learn in two. When I worked in the emergency room, I was on call for 24 hours and off for 24. It was rare when I was on call that I would get any sleep during the night. The ER was usually packed until the wee hours of the morning, especially on Saturday night. Normally I would get off at eight in the morning, go to the cafeteria and eat breakfast, then go to my room and go to bed. I would wake up about four o'clock in the afternoon, get something to eat, fool around for a couple of hours, and go back to bed. I felt as if I was working all of the time, so I decided to change my routine. I decided that rather than going to bed, I would stay up on my day off and do something fun. It was summertime and there were some beautiful golf courses in Atlanta, so I decided that on my day off I would eat breakfast and go play golf. Green fees were cheap in those days and I had gotten privileges at one of the country clubs through my aunt who lived in Atlanta. It took me a while to adjust to this schedule, but I soon learned to sleep 8 hours out of 48, most days, and I kept that schedule almost every day that summer until I rotated off the emergency room that fall. I started playing golf when I was fourteen at a course owned by my best friend's father. Over the years golf has proven not only to be a great character builder, but a teacher of a philosophy that can be applied to both golf and life. In golf you have to play the ball as it lies, no matter where it lands, and no matter how confident you might be with your game, you can never really be sure where your next shot is going. Sometimes that's the way it is with life too. My first set of clubs had been given to me by my great uncle, O. B. Keeler, who was a sports writer for the *Atlanta Journal,* and the best friend and Boswell of the golfing great Bobby Jones. The clubs that he gave me had been given to him by Bobby Jones. I cherished those clubs and played with them for years. I finally retired them to the attic and got a new set, but every once in a while I will take out one of the clubs, grip it, and take a couple of practice swings, and reminisce about what it was like to play with Bobby Jones' clubs when I was a boy.

It would be hard for today's interns to realize what it was like working at the Grady's in the late 1950s. When I was doing my rotation on the emergency room, I was required to go the medical clinic once a week and see my regular patients even though I had been up all night. I remember one clinic day I had worked all night in the ER and I had to go to the clinic without any sleep. The evening before had been busier than usual and I was exhausted. I got to the clinic and started seeing my patients. I had seen several patients when an old man came in for me to check his blood pressure and listen to his heart. I spoke to him, asked how he was doing, and he said fine. I was so exhausted that I crossed my arms and put my head down on the desk and fell asleep. I don't know how long I slept, maybe 15 or 20 minutes, but when I woke up the old man was still sitting there quietly, waiting for me to finish seeing him.

The work was both exhausting and serious, but sometimes there were episodes of comic relief. I recall an incident in the emergency room with one of the surgeons whose name was Dave. Dave didn't really like working in the emergency room, and he liked even less working with the drunks that the police frequently brought in for us to sober up. After they were sober enough to walk, the police would take them to the Atlanta jail. One night we had a drunk strapped on a gurney off in the corner, and when Dave walked by the drunk he decided that he would have a little fun with him. He turned to the drunk and said, "Say fellow, what are you doing here in the emergency room in Birmingham?" The drunk rose up and protested, "I'm not in Birmingham. I'm in Atlanta." Dave insisted that he had been found at the bus station in Birmingham and had been brought to the emergency room. The drunk got very upset and protested that he was sure he was in Atlanta. A few minutes later I walked by and the drunk grabbed me by the arm, and said, "Doc, you've got to help me." I asked, "What's the matter?" He insisted that he had to get out of there. I asked, "Why?" He rose up tight against his restraints and pointed at Dave and said, "You see that doctor over there," and I said, "Yes." Then in a soft voice so as not to be heard, he said urgently, "I've got to get out of here. That doctor's crazy. He thinks he's in Birmingham."

I finished my internship in June of 1958. I was only 25 years old, and I had two years of obligated military service ahead of me. There was a rule at the time that if you went into the service before you were 26, you had to serve two years on active duty, then an additional four years in the active reserves. If you were over 26, you only had the two years of active duty to serve. I didn't mind the two years of active duty, but I wasn't looking forward to spending four more years in the Reserves. I was afraid that I might go into practice and then get called back on active duty. There was a lot going on in the world at the time, particularly in Europe and Korea. My birthday was the later part of September, so I decided I would go home and get a job, and join one of the branches of the service after my birthday. When I got home that summer I started looking around for something to do. I found that one of the family doctors in the town of Wake Forest was looking for someone to cover his practice while he took his family on a vacation. This was the same town where I had gone to college and wasn't very far from home. The doctor's name was Dr. George Mackie and I drove over and talked to him. He showed me around his office and introduced me to his staff, which consisted of a nurse and his wife. We talked for a while, and I agreed to cover his practice for him while he was on vacation. He wasn't leaving until the following week, so he invited me to spend a week following him around. This proved to be a valuable experience, because over the course of a week I learned a lot about how a family doctor goes about his day, and handles patients, although his schedule was different from most. He would wake up about ten in the morning, eat breakfast and head for the infirmary at the seminary. Wake Forest College had moved to Winston-Salem while I was in medical school and the old campus had become a theological seminary. After seeing patients at the infirmary, he would get into his car and head out on the back roads with his doctor's bag to make house calls on his elderly patients who were homebound. He would check their blood pressure, listen to their heart, give them a B-12 shot, and off he would go to the next patient. He did this all afternoon, until around six o'clock when he would come back to his house where his wife would have supper waiting for him. After supper both he and his wife would go to the office. She was his receptionist and checked patients in and out at the front window. He had a nurse who helped him in the back, and the three of them would see patients until the wee hours of the morning. Sometimes it would be four or five o'clock in the morning before they would see the last patient. Someone told me that this was the first vacation that they had ever remembered Dr. Mackie taking.

Dr. Mackie had a concern for his patients that is all too rare today. His relationship with his patients was personal and uncontaminated by managed care. He didn't have an HMO second guessing him about what his patient needed. His patients trusted him, and he cared for them. I look around me today and I see a health care system that is broken. I see many family doctors who are discouraged and disheartened because someone is always looking over their shoulder and second guessing them. They are constantly afraid that they are going to be sued, so they order all kinds of tests that the patient may not need, because of the fear that they might miss something. They are constantly being told how to practice medicine by the managed care companies. They are told which drugs to use, and which test they can and cannot order. I think that the MBAs who run the managed care companies think that the practice of medicine can be operated on a business model motivated by profit, like a bank or an investment company, but their business model rarely has as one of its primary goals personal concern for the patient.

A Year of Psychiatry

When Dr. Mackie came back from his vacation, I started looking for something else to do until my birthday, after which I anticipated going into the service. I learned that the state psychiatric hospital on the other side of town, Dorothea Dix, named after the great mental health reformer, was looking for a doctor to do physical examinations on the incoming patients. I made an appointment to talk to the administrator, Dr. Walter Sikes, and after the interview he offered me the job, and I took it. I had never thought very much about psychiatry. I had always been curious about what made me tick, and I was interested in learning something that would help me with patients, but the psychiatry that I had been taught up to this point hadn't been very helpful. My expectations were not great about what I might learn while I was there, but my experience there turned out to have a major influence on my thinking, and the type of practice that I would eventually have. Although initially I didn't think I was going to be there that long, I wound up spending almost a year at Dorothea Dix Hospital.

One of the first things I realized was why they hired someone to do the physical exams. Psychiatrists don't like to do physical exams; they don't really like to touch patients. After I had been there for a while, I became bored just doing physical exams and I became curious about what happened to the patients after they were admitted, and what kind of treatment they received. I had been there about a month when my birthday came and I applied to join the Army. The Army told me they didn't need any physicians at that time and to apply later. I decided if I was going to be there for a while, I might as well make the best of it, so I went to Dr. Sikes and asked him if I could be assigned to one of the admitting wards and become a resident and join the rest of the house staff. He agreed, and I began a psychiatric residency that fall that lasted until I finally got a call to go into the Army the following spring.

One evening each week the residents would meet at Dr. Sikes' house and he would lead a discussion on an assignment from one of the psychiatric texts. After going over the assignment, we would begin a discussion that was usually wide open and far ranging. The question that I would frequently ask, which would always lead to a vigorous discussion was, how do we define normal? Where anxiety is concerned, how much anxiety should somebody have? How much is normal? No one ever seemed to come up with a good answer. An anxiety disorder was defined by whether the anxiety interfered with the patient's ability to function, but I was never satisfied with this definition, because in all my training in medicine up to this point I had been taught that there were "normals" in the body's functions, and I felt that an anxiety disorder should somehow be defined as an abnormal amount of anxiety.

One of the first patients assigned to me was a very attractive young housewife who had been admitted because she had tried to commit suicide. She was diagnosed with conversion hysteria and depression. There were no antidepressants in those days, and aside from the shock treatments, the only thing that we had to offer was a sympathetic ear. Each day I would go by her room and pull up a chair and listen as she poured out her story to me. She told me that she had been sexually abused by her father as a child, and her recent marriage was going badly. She said that she had tried to commit suicide because she didn't see any other solution to her problems. After a couple of weeks in the hospital she started getting better, and just before she was discharged she confided in me that if it had not been for me she would have killed herself. My confidence was riding high, until I went over the case with my supervisor, who told me that this was just a typical case of transference, and that the same thing would have happened with any therapist. I felt a little let down.

During the time that I was at Dorothea Dix Hospital I had about 150 patients admitted to me, and of that number only six were committed to the hospital for long term care. After discharge some of the patients came back to the hospital's clinic where I followed them as outpatients. One of the more interesting patients I followed was a social worker with a variety of diagnoses, including alcoholism, depression, and probably some degree of schizophrenia. One of her major problems was her sexual identity. From early childhood she told me she had always wanted to be a male. She said that when she was a little girl she went to one of her uncles and asked him to make her a penis. He told her he couldn't do that, but she kept insisting, so one day he gave in and said he would try. A few days later he brought her a penis that he had made out of a piece of pork rind. Weeks went by in therapy and in a session some time later she related to me a dream that she had had since the previous session. In her dream she had been raped and become pregnant. She carried the pregnancy to term, but when she delivered, it wasn't an infant; it was a litter of pigs. I didn't say anything to her at the time, but I couldn't help but remember Freud's theory of penis envy. Freud had theorized that when little girls see the male penis, they become very jealous and feel that they have been short-changed, but when they grow up and have a baby of their own, the baby becomes a substitute for the penis which relieves them of any envy they might have had growing up. When I told this story to my supervisor he shared it with the rest of the staff, and we all agreed that this incident certainly confirmed one of Freud's theories. I have since come to question much of what Freud taught, but in this particular case he was right on the money.

It was an exciting time to be involved in psychiatry. A revolution was taking place before our eyes that was ushering in a new era in the treatment of mental illness. Though we were still using the shock treatments on some patients, there was a new drug called chlorpromazine, trade name Thorazine, that had recently become available that could treat some the most seriously ill patients. Some patients who had been confined to the back wards, unable to feed or dress themselves, were now able to do those things on their own, and even go to the dining hall for meals. Some were able to talk intelligently to staff members and to other patients. Many of them were able to go home, but unfortunately for many there was no home to go home to. Some were sent out of the hospital without adequate preparations for medications and follow-up, and all too often some of them wound up back in the hospital within a fairly short period of time. Unfortunately, we still see this same pattern today.

With the therapeutic triumph of chlorpromazine there arose a question for which we had no answer. If a chemical, such as chlorpromazine could have such a dramatic effect on the way a patient feels and behaves, how could this be integrated with the Freudian idea that everything we feel is psychological? How is it possible for a drug to have such a dramatic effect on the way an individual thinks, feels and behaves? No one had a clue.

Even though I was fascinated with psychiatry, I knew that my true calling was in family medicine, and when I finally got a call from the Army the next spring I knew that it was time to say goodbye to my patients and friends at the hospital, and even though at the time I thought I was leaving psychiatry behind me for good, my experience at Dorothea Dix Hospital awakened in me a curiosity about the mind and the emotions that has led me ever since to look for answers to the questions that first arose when I was a psychiatric resident.

The U.S. Army and France

In April of 1959 I received a commission as a Captain in the US Army and orders to report to Fort Sam Houston in San Antonio, Texas, for basic training. I packed my bags, said goodbye to Mother and Daddy, and drove to San Antonio. When I arrived at Fort Sam I found that I was one of 70 young doctors who had been sent there for basic training. We were assured that in the two weeks that we were there, we would be taught all that we needed to know about being officers in the United States Army Medical Corps. We were given uniforms and assigned to barracks, and one of the first things they tried to teach us was how to march, but we looked more like rejects from a military comedy. Marching was not one of the courses we were given in medical school. A few days later we were taken to the firing range and given instructions about how to fire a rifle. I already knew how to shoot because I had grown up on the farm and had gone hunting with my father many times, but most of the recruits had never had a gun in their hands. On the day we had to walk a target course, side-by-side, firing live ammunition at targets that popped up from behind bushes, I was a little apprehensive, but fortunately no one got shot.

Several days later we were divided into groups of three, given a map and compass, taken out into the Texas brush country, and told to find our way back into camp, and by the way, to watch out for rattlesnakes. I had been a Boy Scout when I was younger, and I knew a little about reading a map and how to use a compass, so I was appointed the leader of our group. One of the members of the group, a city boy, didn't think that I knew what I was doing, and he took off on his own soon after we got started. My other partner and I followed the map course and made it back to camp and reported in to our instructor. When he asked where the third member of our group was, we told him he had decided to go off on his own and must have gotten lost somewhere out in the brush. The instructor sent a group of men out to where we had last seen him and they found him late that afternoon, wandering in the brush, scared but unharmed.

The only other time I got a little apprehensive was the night we had to crawl across an obstacle course under live fire pushing our rifles ahead of us. Overhead search lights were sweeping from side-to-side, and every few minutes explosions would go off somewhere out on the course. Tracer bullets from machine guns streaked across the night sky like supersonic fireflies. Up in the tower, overlooking the course, an instructor warned us not to stand up. When they gave the signal to start, I crawled across the course like a snake trying to escape a garden hoe. When I got to the safety of the ditch at the other end, I looked to my right and left and saw that only one other soldier had made it across the course faster than me. He and I gave each other "a thumb's up," and we leaned back against the ditch and waited for the others to make it across. Out on the course, bombs periodically lit up the night ski, and the tracers kept zinging overhead, and up in the tower the instructor kept shouting encouragement over the loudspeaker to those still out on the course. One of the recruits got turned around and started going back the wrong way, and the instructors kept trying to turn him around and head him back in the right direction. I think it was the same guy from the city who had gotten lost in the brush. I couldn't help but feel sorry for him.

We had been there about a week when we received instructions to go to one of the meeting rooms on the base. When we got there a Major handed us some forms to fill out. On the forms we were to put down where we wanted to spend the next two years. I hadn't given it much thought because I had always assumed that we wouldn't be given any say so about where we were to be assigned. I had assumed the worst and was sure I would be assigned to either Korea or Fort Jackson, SC. I thought about it for a few minutes and I initially put down Germany. Hedrick is an old German name, and I thought I might like to go back to the old country, but the longer I thought about it the more I felt that Germany might not be the right choice. I reasoned in those few short minutes that the German culture might be too regimented and scientific for me after 23 years of school and training, and I wasn't sure that that was the kind of culture I wanted. I thought maybe I should request somewhere where I could learn something about the art of living. I immediately thought of France. That's where

I should ask to be assigned, I said to myself. Though I had little hope that my request would be granted, I erased out Germany and put down France. I could always hope. When the Major came back around at the end of the session and took up the forms, I pointed out to him that I had changed my original request. I told him if it was possible I would like to be assigned to France, anywhere in France, I said. He said he would see what he could do. I wasn't very optimistic, but when my orders came a week later, I had been assigned to the 34th General Hospital, in the village of La Chapelle, just outside of the town of Orleans, about seventy miles south of Paris on the Loire River. Orleans was the headquarters for most of the US troops in France.

Visions of wine, women, and song danced in my head. I had never been to Europe, and only knew about France and Paris from what I had read or seen in the movies. I kept hoping that I wasn't dreaming. I was afraid that I was going to wake up and find out that they had made a terrible mistake and I really was going to the DMZ instead. When the two weeks of basic training was over, and I was driving home, the reality began to sink in I really was going to France.

Two Years in France

When I got home I spent a few days with Mother and Daddy before leaving for France. They weren't at all sure about my being so far from home, and I got the usual parental warning about taking care of myself. I assured them that I would be fine, and promised that I would write every few days. I caught the bus from home to McGuire Air Force Base in New Jersey, and flew with a group of GIs to France in a Lockheed Constellation, a four-engine prop plane, no jets for GIs in those days. We flew over Newfoundland and landed in Iceland, where the plane refueled; and from there across the North Atlantic, over Scotland, and on to France, where we landed at Orly Field just outside Paris.

It was the first day of May of 1959. I was 26 years old, and though I felt wise in the ways of medicine, I felt very naïve in the ways of the world and the art of living, but I was ready to learn. I soon found out that I had come to the right place. During the two years that I lived in France I came to understand why the French are sometimes accused of being a bit conceited about their country. France is truly a beautiful country, and the French appreciate the beauty and the bounty of their countryside, and in the two years I lived there, I came to love and appreciate it too.

It was the spring of the year, and as I rode along in the Army bus from Paris to Orleans, I saw fields green with spring wheat, and flowers blooming everywhere. It seemed that every spare foot of ground not planted in crops, was planted in flowers. There were gardens on almost every corner, and in every window there were boxes filled with flowers. The whole countryside seemed perfumed with the scent of blooms. I imaged that heaven must be a little like this. I was taken to the Bachelors Officer's Quarters and assigned a room. Behind the BOQ was a little river called the Loiret, which flowed into the Loire a few miles downstream. The Loire is the major river that flows westward through the middle of France on its way to the Atlantic. Not too far from the BOQ was a small chateau that had been converted into headquarters for some of the general officers, and nearby was the officer's club which had a bar and a restaurant. As convenient as all of this was, I knew that this was not where I wanted to live for the next two years. Staying there for me would have been like living in a motel and I made a vow that as soon as I could, I would find a place in the countryside and move out.

A couple of days later I was given my duty assignment and I learned that I was to be assigned to the general medical clinic at the hospital where I would see Army dependents and GIs. Later on that summer when the head of psychiatry went on leave, I was temporarily put in charge of the psychiatric unit at the hospital. My duty station changed again midway of my tour and I was appointed the Commanding Officer of the 760th Medical Detachment, which was the dispensary at base headquarters, where my duties were to take care of the general officers assigned to headquarters.

One afternoon, not long after arriving in Orleans I came back to the B.O.Q. and walked over to the officer's club for a drink. I sat down by another Captain at the bar, who I noticed was also a medical officer, and we introduced ourselves. His name was Don Sperling and he worked at the pediatric clinic where he saw the children of dependents. We started talking about where he lived and he told me that he lived in the village of St. Preve-St. Mesmin, which was on the banks of the Loire River across from downtown Orleans. I told him

I was looking for a place to move to, away from the BOQ, and he invited me to come and see where he lived. We got into his white Mercedes 190SL sports car with red upholstery and drove to the village of St. Preve-St. Mesmin. We turned off the main road in the village and came to a stop at the end of a narrow driveway. On each side of the driveway was a high stucco wall. We got out of the car, went through a door in the wall, and stepped into a beautiful garden, with a lawn surrounded by irises, peonies, and roses climbing an arbor. On the other side of the garden was a large house where the owners lived. We walked down a graveled path to a little house of stucco and beams, built like the houses in Normandy. It had French doors that opened into the garden, and in the ceiling was a sky light. He told me it had once been an artist's studio. We walked inside and there was a large main room with a bed, an easy chair, a dresser, an armoire, and a table with four chairs. Off the main room there was a little kitchen with a small coal furnace that fed radiators for heat in winter, and opposite the kitchen there was a small bath. Don told me that every morning after he left for work, the lady who lived next door came over to clean the house and make up the bed for five dollars a week. If I had tried to picture in my mind the ideal place to live while I was in France, I couldn't have pictured a more perfect place.

On the way back to the officer's club, I asked Don if there was any way that I might be able to rent the house after he vacated it. He said he didn't know, but he suggested I stay there while he was away on leave. He was leaving the next day for the South of France with his French girlfriend, and would be gone for two weeks. He told me as soon as he got back, he would be there just long enough to pack, and then would be shipping out for the States. I told him I would like to move in even if it turned out to be only for two weeks. We drove back to the BOQ and I started packing. Don left the next morning, and I moved in that afternoon. I placed my bags on the floor and went outside to look around. In a few minutes a very attractive, petite lady with a charming smile, walked over from the main house and introduced herself. She was the wife of the owner, Madame Giselle Barbauchon, and she proudly gave me a tour of the garden, speaking to me in her best English and I responding in my very elementary French. I had taken Latin in high school and college, and when I arrived in France I didn't know the difference between, "bon jour" and "bon soire," but I was learning, and she told me that my French accent was very good. I could see that the garden was her true passion, and I complemented her on having one of the prettiest gardens I had ever seen. At the end of the tour she asked me if I could come to dinner that evening with her and her husband, and I accepted with both surprise and delight. I later found out that it was somewhat unusual for the French to invite someone to dinner so soon after meeting them, as they usually like to get to know someone better before inviting them into their home. At dinner that evening my French improved as the evening progressed. It was aided in no small part by a Scotch high-ball before dinner, wine with dinner, and an apple brandy from Normandy, called Calvados, after dinner. We talked about my family back home and where I lived in the States. I found out that they had two sons, and they owned a clothing store in downtown Orleans. One of their sons was married and had a young son. I thought to myself, she didn't look old enough to be a grandmother. She looked as though she could only be in her late twenties.

For the next two weeks I would go to the hospital in the morning to work at the clinic, and come back to the little house in the afternoon to enjoy the peace and quiet of the garden. On the table would be a fresh bouquet of flowers Madame Barbauchon had cut for me. Sometimes Monsieur Monseau, the gardener, would still be working when I got home, and I would invite him in for a drink and ask him to tell me about Orleans. He told me in French, as he spoke no English, about the time that General Patton came through Orleans during World War II. That had been some 15 years before, but he spoke about it as if it were yesterday.

Not long after I arrived in France, I was invited by a group of new French friends to go into town and watch the parade of the Festival of Joan of Arc, or *Le Fete de Jean d'Arc*. On the day of the parade I joined my new-found friends in an apartment that overlooked the main square of the city, which was called Le Matrois. We sat on a balcony and drank a white wine from the Loire Valley called Sancerre, ate fresh strawberry tarts, and watched the regional bands dressed in their native costumes march by from the different parts of France. The climax of the festival was a speech by the President of France, General Charles De Gaulle. He spoke from the balcony of the City Hall, which was across the square from where we were watching the festivities. I didn't understand anything that he said, but he sounded very impressive with a deep bass voice. In France he was called "Le Grande Charles."

Don came back two weeks later, and I found out that if I wanted to stay in the house I would have to go to the Franco-American housing authority in downtown Orleans and apply for it. I also learned that the housing authority assigned houses according to rank and length of time in France. I dreaded going because I was afraid that since I was only a newly arrived Captain, some Colonel who outranked me would get the house. I screwed up my courage and finally went downtown. I walked up a long flight of stairs to an office on the second floor and introduced myself to a pretty, young French secretary sitting behind a small desk. I introduced myself and told her why I was there. I told her about the little garden house, and how much I liked it, and how much I would like to rent it now that Captain Sperling was leaving. I kept going on and on about the house when suddenly she stopped me and looked up. She said for me not to worry. She said that it had already been decided that I could have the house. She was in charge of assigning houses, and this particular house belonged to her mother-in-law, and she said her mother-in-law liked me very much. I thanked her, put down my deposit, and floated back down the stairs, very grateful for my good fortune.

I lived in the little garden house for the next two years. I learned to speak French and by the time I left I could pass for French when I went to another country. It was a period that changed my life in many ways. It was a far cry from the life that I had known growing up on the farm and it was a good time to be in France. The country was recovering from the disaster of World War II and the French were regaining their self-confidence. In Paris there was an intellectual excitement and freedom of ideas that I found pretty heady for a young man who had never been far from home. Paris was an hour away on the train, and the movable feast that Hemingway had written about was there for the taking. I frequented the Café Deux Margot where Jean-Paul Sartre was a regular, saw *Carmen* at the Paris Opera, and heard Herbert Von Karajan conduct the Berlin Philharmonic in Beethoven's Ninth Symphony. Though I occasionally dined at one of Paris' fine restaurants, most of the time when I was in Paris or Orleans, I usually ate at one of the neighborhood bistros, which were more my style. One afternoon in the spring of 1961 I was walking around Paris with a girlfriend, and we looked up on the marquee at the Olympia Theater and saw that Edith Piaf was performing inside. The show was almost over and the doors were open. We walked in and stood at the back of the theater. She was all alone on stage, framed by a spotlight, and behind her, mostly hidden by a gossamer curtain, was the orchestra. Chills ran up and down my spine as the little Parisian sparrow sang her heart out. She closed the concert with one of her signature songs, *Non, je ne regrette rien.* No, I regret nothing, a song which made her a favorite of the Existentialists, including Sartre. I later found out that she was deathly sick at the time, and this was one of her last concerts. She died two years later in 1963.

The two years that I spent in France for me was also a period of great reflection. I had been too busy in college, medical school, and internship to do much reflecting on life. During the long winter nights, when it would get dark at four o'clock in the afternoon, I would often come back to my little garden house, put some coal in the furnace, turn the radio to one of the French classical music stations, sit in my easy chair, and read novels that I had never read in school. It had been a long time since I had read just for pleasure, and I found it very different when you read because you want to, and not because you have to. While I was there I read many of the great Russian novels, including Tolstoy's *Anna Karenina,* and Dostoevsky's, *The Brothers Karamazov.* I discovered the great French author Henri Beale, whose pen name was Stendhal, and I fell in love with his books. His most famous was a historical novel called *The Red and the Black*, and after reading it, I liked it so much that I read almost everything that he wrote. I read the classic French tale of seduction *Liaison Dangerous*, after seeing a contemporary French film based on the book. I had a girlfriend who introduced me to the books of the English author Lawrence Durrell, whose most famous works were a series of books called *The Alexandrian Quartet.* The books tell the story of the lives of a group of people in Alexandria, Egypt from four different points of view. I was unable to put them down until I had read them all.

During the two years I was there, I traveled over much of Europe. One New Year's Eve, I went to Rome and drank champagne and danced with the Romans. I went with my Dad to England and Scotland and played golf at St. Andrews. I went to Aintree, England, and saw the Grand National Steeplechase. I traveled to Greece and climbed the Acropolis, and walked the halls of the Parthenon. I skied in Switzerland, Germany, and France, and broke my leg skiing at Kitzbuhel in Austria. I visited a nudist colony on the southwest coast of France, near the village of Montolivet, and learned that the fascination of nudity takes on a new dimension when grandmothers and grandfathers are included. One night I drove up to Paris to meet a date at a little bar on the Left Bank called Le Nuage. I sat down at the bar and ordered a drink and waited for my date to arrive. Sitting next to me was a man a little older than me. He was talking to an attractive young lady next to him and I couldn't help but overhear their conversation. He told her that he was a writer, and he said being a writer was like being an athlete who has to work hard and prepare for the big games. He told her how he had to practice with different words and phrases as he wrote. She listened very intently and finally asked him if he had ever had a book published, and he said yes. She asked him what the name of his book was and he told her it was a book about World War II, and the name of his book was *From Here to Eternity*. I never turned a hair as I listened to James Jones tell his friend about the nuances of writing. I didn't really appreciate what he was saying at the time until I started trying to write myself. For me writing is also hard and demanding. As someone once wrote, "there are no good writers, only good rewriters."

I learned many things in the two years that I lived in France. I learned that even though there are cultural differences between people, there are far more things that we share than separate us. I learned that there are many ways to look at life, and life's problems, and many ways to live life and to solve life's problems. I learned to speak French, and from their language, I learned that there is more than one way to express an idea. The French, for example don't say, "I am afraid;" they say, "I have fear," (*Je peur),* which influenced my way of thinking about anxiety. When you *have* something, it's different than when you *are* that thing. I also learned that there are more ways than one to look at a problem, and how to think outside the box. More than anything I learned about myself. I could have stayed on in France and I thought about it. I had met a number of expatriates living in Paris, but I saw in their lifestyle an ever ascending spiral of sensual pleasures; and I knew that that upward spiral couldn't ascend forever, and sooner or later it would come crashing down. My father's health was declining, and in one of his rare letters, he begged me to come home. Mother was having a hard time and I decided that it was time for me to return home. It was almost two years to the day from when I had arrived in France. I had extended my tour of duty for three months and wasn't supposed to get out of the Army until later that summer, but I decided to cut short my tour of duty in France. I went to my commanding officer and asked for an emergency transfer back to the States to help take care of my father. My transfer was granted, and I was reassigned to Womack General Army Hospital at Fort Bragg, back in North Carolina, and only an hour away from home.

I left France with mixed emotions. It seemed more like home to me now and it was a home I had come to love. When I had traveled to other countries in Europe, even England, whenever I came back to France I felt like I was coming home. I had bought a French sports car called a Facellia, and one of the last things I did before leaving for home was to take the Facellia to St. Nazaire, a port on the west coast of France, where I left it for the Army to ship back to the States. A few days later, I said goodbye to my French and American friends, took the train to Paris, and boarded a plane for home with a heavy heart.

Home From France and Back to Grady

On the flight home I sat by a window and stared for a long time at the ocean below. I thought about what one of my heroes, Thomas Jefferson, had said about France;

So ask the traveled inhabitant of any nation, in what country on earth would you rather live? Certainly in my own, where are all my friends, my relations and the earliest and sweetest affections and recollections of my life. Which would be your second choice? France.

I wondered how Jefferson must have felt when he sailed for America that last time and watched France sink out of sight below the horizon. I had fallen in love with France too. I hoped that one day I could come back, but I knew that when I did, it would never be the same. What I had experienced in the past two years was gone forever and existed now only in my memory.

I still had six weeks of active duty to pull before I could be released from the Army, so when things settled down at home I returned to duty at Womack General Army Hospital at Fort Bragg, where I could come home on the weekends and even during the week if needed. The next six weeks went by rapidly, and when I finished my tour of duty, I resigned my commission and opted out of the reserves. There was a lot going on in the world in the summer of 1961. The

Cold War was heating up and I was not anxious to be called back to active duty after starting a practice. I wanted to get some additional training in pediatrics before I hung out my shingle, so I contacted Grady where I had done my internship, and asked if I could come back and join their pediatric residency program. They said yes, and on July I, 1961, I returned to Atlanta to begin life as a resident in a big city hospital making a $125 dollars per month, and living in a single room at the hospital with a bed, a desk, and a dresser.

It didn't take long for me to grow tired of living at the hospital, so I got a job moonlighting on the weekends in the emergency room of a small hospital north of Atlanta, and with the extra money I started looking for a furnished apartment I could retreat to when I wasn't on call. I found one in an old section of Atlanta which was attractively furnished with antiques and wasn't too far from the hospital. The apartment was owned by an elderly lady who lived in one of the apartments in the same complex, and I went by and introduced myself. I told her I had just come back to Atlanta from France and was a resident at Grady Hospital. I told her how unhappy I was with the room at the hospital, especially after being in France for two years. I told her I wasn't making much money and I wondered if she had anything that I could afford to rent. As luck would have it she had lived in Paris and spoke fluent French. We spent the rest of the afternoon speaking French and reminiscing about our days in France and Paris. Before I left she offered to rent me an apartment at a rate that I could afford, which was considerably less than what she usually charged. My luck was holding out at finding a place to live. I moved in shortly thereafter, and every few days I would go by and visit with her. We would speak French and reminisce about our glory days in Paris.

When I got back to Grady I was glad to see my old professors, especially Dr. Willis Hurst, who had been my cardiology professor when I was an intern. Even though he hadn't seen me in three years, I was very flattered that he remembered me and asked if he could join me for breakfast one morning in the hospital cafeteria. Over breakfast that morning he told me about a trip that he had recently made to Africa with his most famous patient, President Lyndon Johnson. I had always admired Dr. Hurst, even though some of the medical students from Emory were somewhat intimidated by him. I don't think that I ever met a teacher who liked to teach anymore than he did. It was not uncommon for him to come to the hospital on a cold rainy wintry Sunday afternoon, go up on the wards and gather up a half dozen of the residents and interns, take us off to one of the classrooms, sit us down and say, "O.K., boys, what do you know about aortic stenosis?" By the time he let us go, we knew all there was to know about that particular heart value problem.

I spent that fall at Grady on the newborn service and pediatric cardiology. It didn't take long for me to get back into the old routine, but back home things were not going so well. The stress of trying to hold things together was getting to Mother, and in October of that year she was taken to a new hospital that had just opened in Raleigh for emergency surgery on a bleeding ulcer. I flew home to be with her and the family and by the time I got there she was out of surgery, and fortunately was doing well. While I was there I met some of the local physicians and one of them told me that they were looking for someone to work in the emergency room, and he asked if I might be interested. I told him not at that time, but I might consider it when I finished my residency. After I was sure that Mother was o.k. I flew back to Atlanta.

While I was at home it was apparent that Daddy's health was deteriorating rapidly. He had high blood pressure and kidney failure. Complications from cataract surgery had left him with poor vision and he seemed down a good deal of the time. On the way back to Atlanta I wondered whether or not I was going to be able to finish out the year. I was working with a pediatric cardiologist, Dr. Katherine Edwards, who was doing research on congenital heart disease, and I was her first assistant. It was exciting work and though I found the academic atmosphere attractive, I knew that my heart belonged to family practice. I wanted to be involved with the whole patient and to treat all of their ills. Just before Christmas of that year things at home got worse. Daddy's eyes sight continued to fail and Mother was having a hard time keeping his office open. I decided it was time for me to go home. I went to the program director and told him that I hated to leave, but I needed to go home to be with my family. He said he was sorry I had to go, but understood and wished me well.

Returning Home and Starting a Practice

A few days before Christmas of 1961, I put my few belongings in the Facellia and drove home. When I got home I moved back into my old upstairs bedroom. Mother and Dad were overjoyed and relieved that I was home. After a couple of weeks when things had settled down, I started looking for something to do. I remembered that the hospital where mother had been operated on had been looking for someone to work in the emergency room, and I wondered whether the position had been filled. I called the head of the medical staff and asked him if the emergency room position was still open, and he said that it was. I made an appointment to meet with him, and after the interview I got the job. The hospital had only been open for about a year and although the emergency room wasn't very busy, the hospital's charity service was very busy and it occurred to me that I could continue my training while covering the emergency room. I went to the Chiefs of Medicine and Pediatrics and asked if I could continue my training as a resident. They agreed and we worked out a schedule whereby I could continue my training under their supervision while also covering the ER.

For the next six months I worked on pediatrics and internal medicine while covering the ER, and my experience at the hospital proved to be valuable in many ways. I learned the difference between how doctors in private practice handle cases as compared to the doctors in the "Ivory Tower." The local doctors were very helpful in giving me tips about how they treat patients in the real world. I found out that sometimes it's not practical to order every available test on every patient as we used to do when I was in training. It would be prohibitively expensive to practice medicine that way, away from a medical center, and I learned to compromise and order what was necessary to give the patient good, but not wasteful care. My experience at the hospital also gave me a chance to meet the local physicians and see which ones I felt comfortable calling on as consultants when I finally did go into practice.

When I finished my stint on medicine and pediatrics that summer, I felt I was ready to hang out my shingle. In the late 1950s and early 1960s it was not uncommon for a medical student to finish medical school, take one year of internship, and go directly into general practice.

In the late 1960s things began to change. The trend toward specialization was putting general practice in real jeopardy, and the leaders of general practice realized that unless something drastic was done, the family doctor, who usually was a general practitioner, ran the real risk of becoming extinct. In the late 1960s, under the leadership of the American Academy of General Practice, the name was changed to the American Academy of Family Practice, and residencies to train family physicians were established. Instead of one year of internship, the training was expanded to a three-year residency, with a board examination at the end of training. I had put together my own family practice residency before the formal programs were established. My training had consisted of nine months on internal medicine and cardiology, six months on pediatrics, three months on Ob-Gyn, nine months on psychiatry, nine months in the emergency room, and an additional two years in the Army doing general medicine and psychiatry. I felt ready to go into practice. When board exams in Family Practice were first offered in 1971, I took them and became Board Certified in Family Practice. Since that time I have taken a recertification exam every seven years in order to maintain my board certification up until the one I missed in 2005.

On July 2, 1962, I hung out my shingle and began practicing family medicine in my father's old office on the land where I had grown up. Directly behind the office was the garden spot where my brother and I picked butterbeans when we were boys. Most physicians at the time had their offices either downtown in one of the professional buildings, or close to one of the hospitals. I was one of the first to open an office in the suburbs. When I was growing up this had been a rural area, but it was rapidly becoming suburban. The two-lane road in front of the office had become a four-lane thoroughfare, and there was a growing housing development in the woods behind the house where my brother and I used to ride our horses. My Dad's old office was not very big. It had a waiting room and two examining rooms. In one of the exam rooms I also had my office, and in the other exam room I had a small lab off in one corner where I did a limited amount of lab work. I bought three used wooden examining tables from the hospital, which I still have some 40-plus years later. I bought a used EKG machine and I used my old medical school microscope in the lab.

When I first opened my office, a routine office visit was four dollars, and most patients paid cash. When I began practice I covered both the emergency room and my office, going back and forth during the day. After six months the practice had gotten busy and I gave up the emergency room. The patients who came to the office were from all walks of life. Some came from the emergency room, others from the area around the office. Some were neighbors that I had known since I was a boy, and still others from a new housing development that was being built in the woods behind the office. I even had some of my old high school classmates come, which I thought took a lot of courage on their part and made me feel very proud. I saw tenant farmers who lived in houses heated by a wood stove, whose source of water was a well behind the house, and whose toilet was down a path. I had attorneys come from an upscale housing development that was being built in one of the fields where we used to play baseball when I was a boy. Over the years I have had a number of physicians and their families as patients, and on one occasion a psychiatrist.

This was the kind of practice I wanted. I didn't want to see just one kind of patient, or treat one kind of illness, I wanted to see all kinds of patients and treat all kinds of illnesses, and over the past 40-plus years I have never gotten bored with what I do. Even though managed care makes my life more complicated now, I am still glad to get up and go to the office every morning.

The Search for a Solution to a Vexing Problem

The patients who came to the office had a wide variety of problems, heart disease, diabetes, thyroid disease; and though I felt competent to take care of most of these, there was one group of patients I found especially challenging, and it wasn't the patients with complicated heart disease, or difficult to treat diabetes, it was the patients with emotional problems. Even though I had had almost a year of training in psychiatry, I still found myself struggling to deal with the many emotional problems that patients had. So I began looking for a better way to understand and treat one of the most vexing set of problems the average physician faces. Psychoanalysis was the only psychotherapy we had at the time, but it wasn't practical for the family doctor's office, and though we had a new set of anti-depressants and anti-anxiety drugs, some of them had significant side effects and were often ineffective.

In the 1960s and 70s several new forms of psychotherapy were developed that were more practical and easier to use than psychoanalysis. One that became very popular was called transactional analysis. It resembled Freud's model of the psyche, but with a modern twist. In TA, as it was called, the superego became the parent, the id became the child, and the ego became the adult. The change in terms made it easier for the average person to understand and apply to their lives, but transactional analysis was more helpful in interpersonal relationships than it was with emotional problems like depression and anxiety. There were also other types of psychotherapy that became popular in the 1970s, including group therapy, couples therapy, gestalt therapy, and 12-step therapy. These new therapies were primarily based on outcome and utility, and emotional problems like depression and anxiety were a secondary concern.

Transactional analysis, like psychoanalysis, depended primarily on insight to help patients; while many of the other therapies used the dynamics of group interaction to affect change. There was one new therapy that made more sense to me than any of the others, it was called cognitive therapy. Cognitive therapy was based on the theory that if you could change the way people see things, you could change the way they feel and behave. There were several reasons why cognitive therapy appealed to me. For one, it seemed to be more effective than the other forms of psychotherapy, and patients seemed to understand it better. It was also more reasonable and rational than other therapies; in fact, it was called rational emotive therapy by one of its founders and leading practitioners, Dr. Albert Ellis, who was a psychologist, not a psychiatrist. He was fond of quoting the Stoic philosopher, Epictetus, who had written many centuries before:

People are disturbed not by things, but by the view which they take of them.

Dr. Ellis expanded on this concept and came up with 10 irrational ideas which he felt were the primary reasons that people disturb themselves, and which led them to having unhealthy emotions and behaviors. He reasoned that if the therapist could get the patient to challenge their irrational ideas and beliefs, and change them, the patients would not only feel better, but also behave better. He developed the formula A + B = C, where A stands for the activating event or life experience, B stands for the patient's belief system, and C stands for the subsequent emotional response and behavior. He concluded that it was the belief system that held the key to emotional health, and if the belief system was correct and realistic, then the emotional response would also be correct; but if the belief system was faulty, then the emotional response would also be faulty.

Dr. Ellis' 10 Irrational Ideas.

Number one is the idea that you must, yes must, have love and approval from all the people you find significant.

Number two is the idea that you must prove thoroughly competent, adequate, and achieving, or a slightly saner version, that you must have competence or talent in some important area.

Number three is the idea that when people act obnoxiously and unfairly, you should blame and damn them, and see them as a bad, wicked, or rotten individual. And the corollary to this, of course, is that when we act that way we should also be blamed or damned.

Number four is the idea that you have to view things as awful, terrible, horrible, and catastrophic when you get seriously frustrated, treated unfairly, or rejected.

Number five is the idea that emotional misery comes from external pressures and that you have little ability to control or change feelings.

Number six is the idea that if something seems dangerous or fearsome, you must preoccupy yourself with and make yourself anxious about it.

Number seven is the idea that you can more easily avoid facing many life difficulties and self-responsibilities than undertaking more rewarding forms of self-discipline.

Number eight is the idea that your past remains all-important and that because something once strongly influenced your life, it has to keep determining your feeling and behavior.

Number nine is the idea that people and things should turn out better than they do and that you must view it as awful and horrible if you do not find good solutions to life's grim realities.

Number ten is the ideas that you can achieve maximum happiness by inertia or inaction or by passively and uncommittedly "enjoying yourself."

Dr. Ellis was a very prolific writer who had published a number of books on rational emotive therapy, including a very popular one entitled, *A Guide to Rational Living*. He had an institute in New York dedicated to his ideas where weekly workshops and seminars were held. I had read a number of books and articles by him, and in the spring of 1975 I drove to Chicago to attend one of his five day workshops on RET.

When I got to the conference I found that I was one of only a small number of physicians who were attending. The vast majority of the attendees were psychologists. Cognitive therapy was relatively new at the time, and there were only a few psychiatrists using it. I knew of Dr. Maxie Maultsby, a psychiatrist from the University of Kentucky, who had spoken at the annual meeting of the North Carolina Academy of Family Practice several years earlier at my request. I was also familiar with the work of Dr. Aaron Beck, a psychiatrist in Philadelphia, who was doing cognitive therapy in the treatment of depression.

Dr. Ellis was a tall thin man with piercing eyes who smiled easily. He spoke with a great deal of passion and reminded me of a preacher, and from the enthusiasm of his audience it was apparent that he had a lot of converts. I was very impressed with what he had to say and his brand of therapy made more sense to me than any of the other therapies that I had studied.

Arriving at a Definition of Normal

The conference was composed of a series of lectures and workshops, and it was in the workshop on the ten irrational ideas that I had something of an epiphany. As I listened to the lecturer, it suddenly occurred to me that there were not ten different errors in thinking, there was only one error, and that was an error in the *weighting* of the significance of an event, or circumstance, by the individual, and I realized that I could use this weighting mechanism to define a normal emotion.

If we assume that the data coming into the brain about a circumstance is accurate, and the weighting of the significance of that data by cognition is also accurate, it follows that a <u>normal</u> emotional response would be that response in which the type and amount of the emotion, or emotions, matches and is proportional to the circumstance. On the other hand, when cognition is off, and the weighting of the significance of the circumstances is not accurate, a normal emotional response is not possible.

Defining normal in psychiatry has always been elusive, but I felt for the first time since I started looking for a definition of normal when I was a resident, I had arrived at a definition of normal that was reasonable and rational. Sometime later I read that Aristotle had defined normalcy essentially the same way, when he was writing about sadness and depression. He said the sadness someone feels is normal when it is proportional to the loss, and defined depression as being sadness without cause, a concept that has come down through the psychiatric literature ever since, but the concept that depression is sadness without cause is one that I disagree with, as the reader will find out later on in the book, for I believe that depression and sadness are two entirely different emotions and though they may occur together they are not the same.

Since we don't actually have the means to weigh either the significance of an event or the amount of an emotion, for practical purposes an arbitrary scale of one to ten can be used to measure both.

If, for example, an event will have little impact on an individual, a level one on a scale of one to ten, then the normal emotional response to that event should also be a level one. On the other hand, if the impact of an event is great, a level ten, then the normal emotional response should be a level ten. Some day in the future I am convinced we will be able to measure the amount of an emotion in the lab, but for now the one to ten scales will have to do, and actually works pretty well in practice. I have also found that patients can be taught to use a one to ten scales to measure their own emotions, which has the added benefit of helping patients gain a sense of control over their emotions which they may never have had before.

If someone is in great danger, a high level of anxiety is normal. Imagine you are walking along the edge of a cliff, the Grand Canyon for example, and the path suddenly gives way beneath you. You reach out and grab a limb as the path collapses. It is mile to the bottom of the canyon. Under a circumstance such as this, a very high level of anxiety would be normal, a level ten. But if someone has the same amount of anxiety while standing in line at the grocery store, that would not be normal, because the amount of anxiety would not match the circumstances. When someone has a level ten of anxiety in the grocery store, or a movie theater, or as happened to one of my patients when he was in the barber's chair, it is called it a panic attack.

But let's go back to the grocery store for a moment. How much anxiety should someone have at a grocery store, how much is normal? Well, it depends. Let's change the circumstance of the visit to the grocery store a bit. It is late at night in a crime-ridden neighborhood, and someone has to go to the only grocery store open to get some milk for the baby, a grocery store where there have been multiple holdups and several murders. Just as the individual reaches the checkout line, two men whose faces are covered with ski masks, walk through the door with guns in their hands. How much anxiety would be normal under this circumstance? It would certainly be much higher than a circumstance where the grocery store was just down the street from where someone lived, in a safe neighborhood, and in a store that had never experienced any crime.

So, without taking into consideration the circumstance or context under which the anxiety, or any emotion occurs, it is impossible to say whether the type of the emotion, or the amount of the emotion is normal or abnormal. But it is possible to say when an emotion, and the amount of the emotion is normal, if the circumstances are included in the equation. The DSM leaves out both of these factors, and as a result often comes across as being very vague and uncertain. In the DSM a disorder is based primarily on the individual's symptoms, whether the individual is able to function, and/or whether the emotion is causing the individual any discomfort. It is often left up to the therapist to guess whether an individual's anxiety is normal or not. There are three questions about the emotions the DSM basically ignores. The first is the context in which the emotion occurs. The second is whether the emotion is proportional to the context. The third is a cause for an abnormal emotion. Without an answer to all three of these questions, arriving at a definition of a normal or an abnormal emotion is impossible.

In psychology and psychiatry the word normal has always been something of an anathema. The word normal comes from an old Latin word, *norm*, which means a carpenter's square, and it first appeared in the English language sometime before 1500. In medicine we know how to measure thousands of the chemistries of the body's internal environment, and we know what the normal range is for them. But what about the chemistries that help us survive in and relate to the external environment? Darwin said that the emotions were built in to help us survive in the external environment, and if Darwin is right and the emotions are those chemicals that help us survive in the external environment, a normal emotional response would be that response where the amount and type of a survival emotion matches the circumstances, and falls within a normal range, just like the blood sugar and thyroid hormone. The DSM unfortunately fails to take these concepts into account. As an example of its vagueness the DSM list the following criteria for the diagnosis of a generalize anxiety disorder. (DSM-IV fourth edition)

A.   Excessive anxiety and worry (apprehensive expectation), occurring more days than not for at least 6 months, about a number of events or activities (such as work or school performance).

B.   The person finds it difficult to control the worry.

C.   The anxiety and worry are associated with three (or more) of the following six symptoms (with at least some symptoms present for more days than not for the past 6 months). Note: Only one item is required for children.

    (1)   restlessness or feeling keyed up or on edge
    (2)   being easily fatigued
    (3)   difficulty concentrating or mind going blank
    (4)   irritability
    (5)   muscle tension
    (6)   sleep disturbance (difficulty falling or staying asleep or restless unsatisfying sleep)

D.   The focus of the anxiety and worry is not confined to features of a...panic attack, being embarrassed in public, ...being contaminated,...being away from home or close relative,...gaining weight,...having multiple physical complaints,...or having a serious illness,...and the anxiety and worry do occur exclusively during Posttraumatic Stress Disorder.

E.   The anxiety, worry or physical symptoms cause clinically significant distress or impairment in social, occupational, or other important areas of functioning.

F.   The disturbance is not due to the direct physiological effects of a substance...or a general medical condition...and does not occur exclusively during a Mood Disorder, a Psychotic Disorder, or a Pervasive Developmental Disorder.

Nowhere in the description of generalize anxiety does the DSM ever say anything about the circumstances, or how much anxiety the patient has, and without knowing the answer to both of those questions, it becomes impossible to say whether the amount of anxiety the patient has is normal or not, and by extension whether they actually have generalize anxiety disorder.

Different Theories on the Emotions

One of the major problems in arriving at a rational theory of the emotions is the general confusion that exists about the emotions. Where do they come from, what is their significance, which of the emotions are primary or basic, and which ones are secondary? Over the years philosophers, physicians, and psychologists have generated many theories about the emotions. One of the first was Aristotle, who in his second book of the Rhetoric listed the following emotions:

Anger: *An impulse to revenge that shall be evident, and caused by an obvious, unjustified slight with respect to the individual or his friends. Slights have three species: contempt, spite, and insolence.*

Mildness: *The settling down and quieting of anger.*

Love: *Wishing for a person those things which you consider to be good—wishing them for his sake and not your own---and tending so far as you can to affect them.*

Enmity *(hatred)*: *Whereas anger is excited by offences that concern the individual, enmity may arise without regard to the individual as such. Anger is directed against the individual, hatred is directed against the class as well.*

Fear: *A pain or disturbance arising from a mental image of impending evil of a painful or destructive sort.*

Confidence: *The opposite of fear. Confidence is the hope (anticipation), accompanied by a mental image, of things conducive to safety as being near at hand, while causes of fear seem to be either non-existent or far away.*

Shame   *A pain or disturbance regarding that class of evils, in the present, past, or future, which we think will tend to our discredit.*

Shamelessness: *A certain contempt or indifference regarding the said evils.*

Benevolence: *The emotion toward disinterested kindness in doing or returning good to another or to all others; the same term represents the kind action as an action; or the kind thing done considered as a result.*

Pity:	*A sense of pain at what we take to be an evil of a destructive or painful kind, which befalls one who does not deserve it, which we think we ourselves or some one allied to us might likewise suffer, and when this possibility seems near at hand.*
Indignation:	*A pain at the sight of undeserved good fortune.*
Envy:	*A disturbing pain directed at the good fortune of an equal. The pain is felt not because one desires something, but because the other persons have it.*
Emulation:	*A pain at what we take to be the presence, in the case of persons who are by nature like us, of goods that are desirable and are possible for us to attain---a pain felt, not because the other persons have these goods, but because we do not have them as well.*
Contempt:	*The antithesis of emulation (Persons who are in a position to emulate or to be emulated must tend to feel contempt for those who are subject to any evils [defects and disadvantages] that are opposite to the goods arousing emulation, and to feel it with Respect to these evils).*

Some philosophers have called the emotions the passions, and some of the more rational philosophers tended to favor reason over the passions, and felt that the passions, or emotions, should be suppressed or ignored. Others philosophers, such as Soren Kierkegaard, the Danish Christian philosopher, thought that the passions played an important part in our lives and should be celebrated. One important figure who wrote a great deal about the emotions from a different field was Charles Darwin, who in addition to his works on evolution, was intensely interested in the emotions and wrote a book about them. He felt that the emotions were universal in all living creatures and were not confined solely to mankind. He felt that the emotions had developed over the course of evolution and helped us survive in and relate to the external environment.

In the later part of the nineteenth century, the American philosopher-psychologist at Harvard, William James, and a Danish physiologist by the name of Carl Lange, simultaneously developed a theory of the emotions that became known as the James-Lange theory. It is best described in James' own words.

My theory ... is that the bodily changes follow directly the perception of the exciting fact, and that our feeling of the same changes as they occur is the emotion. Common sense says we lose our fortune, are sorry and weep; we meet a bear, are frightened and run; we are insulted by a rival, and angry and strike. The hypothesis here to be defended says that this order of sequence is incorrect ... and that the more rational statement is that we feel sorry because we cry, angry because we strike, afraid because we tremble ... Without the bodily states following on the perception, the latter would be purely cognitive in form, pale, colorless, destitute of emotional warmth. We might then see the bear, and judge it best to run, receive the insult and deem it right to strike, but we should not actually feel afraid or angry.

James is noted for saying, "I feel sad because I weep, not I weep because I am sad." The problem with the James-Lange theory is it reminds me of the question that we used to ask in college, if a tree falls in the forest and no one hears it, does it make a sound? And the answer depends on how sound is defined. If one defines sound as being solely the presence of sound waves, then the answer is yes, the tree does make a sound when it falls in the forest, even if no one hears it. On the other hand, if one defines sound as requiring the presence of an ear drum for the sound waves to bounce off of, and a tree falls in the forest and no one hears it, the answer is no, it doesn't make a sound. The James-Lange theory of emotions appears to depend on the *brain's interpretation, or awareness,* of the emotional neurotransmitter state as being the emotion, and not the emotional neurotransmitters themselves, and if we apply the sound wave analogy to the emotions, the emotional neurotransmitters would be considered the sound waves, and the brain the ear drum. If the brain is not aware of the emotional neurotransmitters, or fails to recognize the emotional neurotransmitters, or misinterprets the emotional neurotransmitters, even though the emotional neurotransmitters are there, James-Lange would say that no emotion has occurred, or the wrong emotion has occurred. In James' own example, "I feel sad because I weep," he fails to take into account what happens when we weep for joy? The weeping is the same, but certainly the emotion is different. James never addresses this discrepancy as far as I know, and when we weep for joy, it is certainly not the case that we are really sad, but just don't know it.

On the other hand if we assume that the emotions come from specific emotional neurotransmitters, even if the brain fails to recognize the neurotransmitters, or misinterprets the neurotransmitters, the emotion is still there. If we take anxiety as an example, if someone has anxiety neurotransmitters in their brain, the anxiety neurotransmitter are the emotion, and not the brains interpretation of those neurotransmitters that is the emotion. The brain can make a mistake, cognition can be off, and just as when the tree falls in the forest and no one hears it, a sound has still occurred.

One of the next theories about the emotions came from Dr. Walter B. Cannon of homeostasis fame. He, along with one of his Harvard colleagues, Phillip Bard, developed a theory in which they said what happens after an activating event is an arousal in the thalamus, and the emotion and the arousal occur at the same time. We now know there are other areas in the brain beside the thalamus involved in the emotions, but I believe that the Cannon-Bard theory is closer to the truth than the theory of James and Lange. The Cannon-Bard theory assumes that there must be an activating event for an emotion to occur, but their theory fails to explain what happens when an emotion occurs spontaneously, and there is no activating event, such as what happens in individuals with bipolar disease, who have periods of depression and mania totally independent of any activating event.

In the 1960s two psychologists Stanley Schachter and Jerome Singer advanced another theory of the emotions. They felt the human emotional landscape was too varied for us humans to have a particular emotion without an explanation, and that some kind of reason or interpretation had to be applied to any feeling or arousal before the emotion could be identified. In many respects their theory is not unlike the James-Lange theory because in both cases some kind of recognition, evaluation, or interpretation of a particular neurotransmitter state by the upper brain is necessary for the identification of a given emotion.

Richard Lazarus of the University of California advanced a similar theory when he said that before any emotion could be present, there had to be some thought about an event that preceded the emotion. Lazarus' theory was, first the event, then the thought, then the emotion, but again this fails to explain a condition like bipolar disease or panic attacks, where an emotion occurs without there being a thought or activating event that precedes the emotion.

Though many of the early theories about the emotions were advanced by psychologists, one of the latest to propose a theory of the emotions is Dr. Antonio Damasio, a neurologist and physician. He has written a number of books on the emotions, including the one I quoted from earlier, *Looking for Spinoza*. In a critique of Dr. Damasio's work Colin McGinn feels that Dr. Damasio has basically reworked the James-Lange theory and given the brain's perception of the emotional neurotransmitters a new set of names. He calls the brain's, (or the mind's), perceptions of the emotional neurotransmitters, bodily maps, but these maps are not unlike James-Lange's theory, where it is the interpretation of the neurotransmitter state that is the emotion, rather than the emotional neurotransmitters themselves being the emotion.

Yet another theory was advanced several years ago by Dr. Robert Plutchik a psychologist who died in 2006. Dr. Plutchik felt that there were eight basic emotions: joy, sadness, acceptance, disgust, fear, anger, expectation, and surprise; and that each one of these emotions occurred in different degrees or amounts. He, like Darwin, felt the basic emotions developed through the process of evolution, and that each of them had survival value. I agree with much of what Dr. Plutchik says about the emotions, and would include fear, anger, joy, and sadness among the primary emotions, but I would put acceptance, disgust, expectation, and surprise in the general category of feelings, and not include them with the primary emotions. I do agree with Dr. Plutchik that the primary emotions occur in different degrees and amounts.

The basic issue in arriving at a theory of the primary emotions revolves around whether the primary emotions are pure, built in, fundamental, and the same for everyone; or whether the primary emotions are relative and subject to cognition, or the brain's interpretation, as James-Lange and Dr. Damasio seem to imply.

Redefining the Primary Emotions

With so many theories about the emotions and no general agreement about which, if any, are primary, I decided to look anew at what some of the early pioneers in medicine had to say about the body and brain. One of the first places I turned to was the work of the great 19th century French physiologist, Dr. Claude Bernard. He was one of the first to advance a general theory of how the body works. Through his research he came to believe that in health the body has an internal chemical balance that is very stable. He called this the *milieu interieur,* and he felt that it was the stability of our internal chemistry that keeps us healthy.

In the early part of the 20th century, Dr. Bernard's work was taken up by the equally great Harvard physiologist, Dr. Walter B. Cannon, who in addition to continuing Dr. Bernard's work on the *milieu interior,* did extensive work on the emotions and their physiologic consequences, such as the elevation of blood pressure and the pulse rate in anxiety and anger. He expanded on Dr. Bernard's work and wrote a book called *The Wisdom of the Body.* In this book he coined a new term for the *milieu interieur,* he called the process *homeostasis,* which means the same state, a word which soon became a byword in physiology. Dr. Cannon said that our bodies contain multiple systems that are very finely balanced, and though these systems appear stable at the macro level, at the micro level they are very unstable and are constantly shifting back and forth, but always within a normal range when we are healthy. He felt it was the instability at the micro level that enabled the body's internal environment at the macro level to be very stable and give the body great resilience and flexibility.

The mechanism that controls the body's blood sugar is a good example. The body burns (oxides) glucose as its primary fuel source, and the mechanism that controls the burn of this fuel involves multiple systems that work together to keep the blood sugar within a relatively narrow range. Glucose is found in sweets and carbohydrates, but the body can also make glucose from proteins and fat if necessary, a process which takes place in the liver. Insulin, which is produced in the pancreas, controls the conversion of sugar into energy in the cells in much the same way that a damper controls the burn in a wood stove. After we eat a meal, the level of sugar in the blood begins to rise, and the pancreas, when it is working normally, produces insulin. Insulin's function is to open the cell's damper, which lets the sugar burn and causes the sugar level to fall. If the pancreas is unable to put out enough insulin, or there are factors which cause insulin to be less effective (insulin resistance), then the sugar can't burn and the level of sugar in the blood rises. Exercise has much the same effect as insulin and helps open up the cell's damper so that the sugar can burn. Normal blood sugar fluctuates from around 80 to 120, but within that range it can vary widely. As long as the sugar stays within that range, we consider it to be normal, but if it gets very far above that range, we say it is abnormal, and a disease state exist, one that we call diabetes. This is just one example of the thousands of systems that enable our bodies to maintain an internal environment that is stable, and allows most of us to live out our "three score and ten," or better.

A Redefining of the Primary Emotions

Our bodies exist in two worlds, an inside world, or internal environment, and an outside world, or external environment. On the inside of us is a world that for the most part is invisible, and when we are healthy, that internal world is carefully maintained in a homeostatic balance. But we also live in an external world, and in order to survive in, relate to, and stay in homeostatic balance with the outside world, our bodies have developed a set of neurotransmitters which enable us to stay in homeostatic balance with it as well. I believe that what we call the primary emotions are the surviving and relating neurotransmitters which help us stay in homeostatic balance with the external environment.

The word emotion comes from the Latin word *emovere* (e=out and movere=to move), to move out or to stir up. In the 1660s it was used to refer to a strong feeling. Today we generally think of an emotion as being any strong feeling. The words feeling and feelings come from *felen*, an old English word which initially meant to feel or touch something in a physical sense. The first time it was used in the English language to express feelings of a non-physical kind, such as joy or sorrow, was in 1369. Today we use the words *feel* and *feelings* to express not only what it means to feel something in the physical sense, like pain, but also to express feelings of a non-physical nature, like love, anxiety, or disappointment.

I propose that we define *feeling, or feelings,* as being any neurotransmitter state that the brain registers, and that would include the primary emotions, complex feelings such as disgust and embarrassment, and physical feelings such as touch or pain. We feel things physically when the brain registers the neurotransmitters that come to it from the nerve endings in the skin, the muscles, or the stomach. We feel heat and pain when we touch a hot stove, because the nerve endings in the fingers bring a message to the brain from the heat and pain sensors in the fingers, by way of axons and synapses, the wires and connections that link the sensors in the skin to the brain. The brain feels the neurotransmitters of non-physical feelings the same way. We feel anxious when our brains register anxiety neurotransmitters coursing through it. When we feel love, we feel a blend of neurotransmitters that we call love; but it is well to remember that not everyone has the same neurotransmitters coursing through their brain when they use the term love. The same is true of a myriad of other non-physical feelings, such as disgust, shame, or excitement, which are sometimes labeled primary emotions, but I would not put them with the primary emotions as Aristotle and Plutchik did, because I think they are too complex and are not universally the same for everyone.

I propose that we define the primary emotions as those emotions that allow us to survive in and relate to the outside world, and that we divide them into two groups, survival and non-survival. I also propose that each of the primary survival emotions comes from a specific neurotransmitter site located in the lower brain and is unique to each emotion. I believe, like Darwin, that the primary survival emotions developed through evolution, are built in and are universal in all humans, and probably in animals as well. The two non-survival emotions may not be as specific as the survival emotions, but are distinctive nevertheless.

When we are alive there is a constant mix of neurotransmitters coursing through our brains, and when we are conscious and focus on those neurotransmitters, we can become aware of the feelings and emotions that those neurotransmitters represent, but it is well to point out that our emotional and feeling neurotransmitters don't carry any labels with them, and it is only through learning and a general agreement about what we call a particular feeling or emotion are we able to communicate with one another about them. Even though we think we know what someone else is feeling, we need to be aware of the capriciousness of knowing exactly what someone else is feeling. The lack of specificity and the uncertainty about the emotions is one of the reasons there is so much controversy among scholars, doctors, and patients about what constitutes a given emotion. Many times when I see patients, I find they are confused about which emotion they are actually feeling. They confuse anxiety with depression, and vice versa, and I often find myself educating the patient about the emotions in general before we talk about which emotion they have. It is only after I think they understand the different emotions am I able to talk with them about which emotion they are actually feeling, and be able to say whether that emotion is normal or not.

The Six Primary Emotions

Two of the survival emotions are generally thought of as defense mechanisms. The other two are not usually thought of as survival emotions at all, and the last two are unrelated to survival as such, but are the responses to two important events that happen to us as we go through life. All of the primary emotions are paired and opposite.

Probably the easiest of the primary survival emotions to understand is anxiety. Anxiety is the brain's and body's response to the threat of harm or danger. It is the emotion that prepares the individual to seek a safe place, and if it is not possible to escape the harm or danger, it prepares the body to defend or fight. It has often been called the "fight or flight response." We are here today because our ancestors survived by being able to outrun danger or defend themselves against it. The ones who didn't run fast enough, didn't make it to become our ancestors and their genes died out, so it is not surprising there are many individuals around today with very active anxiety centers.

The next primary survival emotion is the opposite of anxiety. While anxiety wants to move us away from something, and helps us survive by taking us out of harm's way, the next survival emotion wants to move us forward or toward something. It is the emotion which enables us to survive by securing or protecting what is necessary for our survival. That emotion is anger-aggression.

The next two primary survival emotions were also built in to help us survive, but are not generally included in lists of the primary emotions. The first of these emotions is depression, and despite what is generally taught that depression is caused by some kind of chemical imbalance, or a form of sadness without cause, I believe that depression was built in as an evolutionary defense mechanism, and even today is normal under some circumstances. Depression is basically a form of hibernation, and when we are truly helpless and hopeless, it is the emotion which helps us survive by conserving resources until things get better. The incidence of normal depression might not be quite as rare as one might think. Since I started looking for instances of normal depression, I have read not infrequently of individuals who have survived long periods of helpless, hopeless, circumstances from earthquakes, avalanches, wars, storms, auto accidents, and other natural disasters. When we are truly helpless and hopeless, our best hope of survival is to enter a state of depression-hibernation which will conserve our resources until we are rescued or things get better.

The opposite of depression, and I wish that we had a better term for it, is mania, which I also believe is one of the survival emotions. I believe that mania is normal when it occurs in the right amount and at the right time. The key to this emotion, as with all of the emotions, is the amount and the circumstance. The manic neurotransmitter is what gives us the energy and the confidence to succeed. We need a little bit of manic neurotransmitter every morning to get the day's work done, but when it is present in excess, such as when someone is in the manic phase of bipolar disease, or when someone stimulates their manic center with a drug like cocaine, a person will feel too confident, better than normal, an emotional state that can be very addictive, and will often get someone into trouble.

The next two primary emotions are often confused with depression and mania, but they are two entirely different emotions. I call them the relating emotions, and they are the normal emotional responses to two major events that happen to all of us as we go through life. The first of these is the emotion we feel when we lose something or someone important to us; and that emotion is sadness, not depression, and the depth and the degree of the sadness we feel should be proportional to the value of whom, or what, is lost. There is a great deal of confusion about the difference between depression and sadness. Even the DSM confuses the two, because it states that one of the symptoms of depression is a sad mood. Sadness is one of the most difficult and painful emotions we have to face in life, but it is also a normal part of life, and one that we all have to face. It is not an emotion that antidepressants can assuage, nor is it due to a chemical imbalance.

The last of the primary emotions is happiness, the emotion opposite sadness, and the emotion we feel when we gain something of value. Just as sadness and depression are not the same, happiness and mania are not the same thing either. The amount of happiness we feel when we gain something of value, is determined by how valuable the gain is to us: the greater the gain, the greater the happiness. The gains that can cause happiness can come from something tangible, such as money or the birth of a child, or from something intangible, such as when *my* team wins the championship.

In the chapters that follow I will start with anxiety as a model for all the primary survival emotions, explain how I think each works, when they are normal and when they are abnormal, and explain how drugs and psychotherapy can both work in the same illness.

Anxiety, the Emotion of Escape from Harm and a Model for the Other Primary Emotions

The one emotion that everyone agrees is one of the primary emotions is anxiety. It is probably universal in all living things and the emotion we feel when we become aware that we might be injured or harmed in some way. It is what prepares us to get away from the harm or danger, and if it is not possible to escape the danger, it prepares us to defend against it. We have all felt anxiety. Our hearts start beating faster, our muscles get tense, there is a feeling of apprehension, and when it is severe we feel like we are going to die.

The words anxiety and anxious come from the Latin word *anxius*, which means choking or distress. It first appeared in the English language in the early 1600s. In the 19th century it was felt anxiety came from the nervous systems, and the term "nerves" became synonymous with anxiety. Even today when we say someone has "bad nerves", it usually means that they have an anxiety problem.

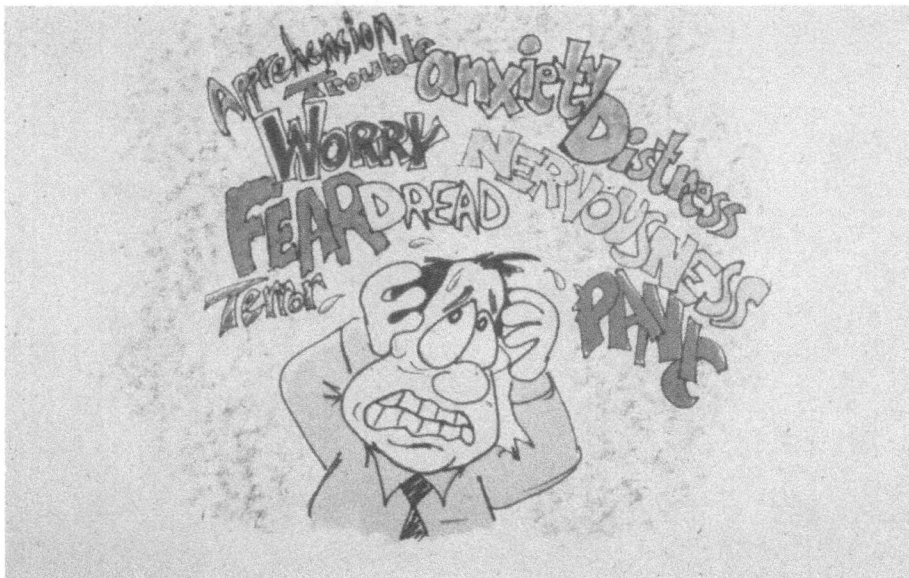

Since we can't draw a blood sample and measure the exact amount of anxiety that someone has, we use words and expressions to convey different amounts of this emotion; words like anxious, apprehension, concern, the creeps, disquieting, distress, dread, butterflies in the stomach, foreboding, nervous, panic, unease, afraid, fear, worried, scared, angst, cold feet, terror, horror, phobia, trepidation, uneasy, consternation, alarm, perturbation, uncertainty, disturbed, frightened, frantic, troubling, unsettling, jittery, jumpy, overwrought, uptight, alarm, frightened, intimidated, and startled.

All of these words express different amounts of the same emotion. Some words express low levels of anxiety, others medium levels, and others high levels; but all are basically expressing the same primary emotion. But what exactly is it that makes us feel anxious? Does the feeling of anxiety arise from outside of us, or from inside of us? The ancients used to think all of our thoughts and feelings came from outside of us. They believed our feelings and thoughts were put there by the gods. Today we know that everything we feel, including our emotions, arises from the neurotransmitters flowing through our brains and bodies. We also know that the mind and the body are not two separate entities, as Plato and Descartes once thought, and when we feel anything, including an emotion, it is because there are neurotransmitters in our brains and bodies that cause us to feel that way.

After the discovery in the 1960s of the anxiety-relieving drugs, the benzodiazepines, an intensive effort was undertaken to understand how they worked, and what neurotransmitters might be involved in anxiety. In 1967 an investigator by the name of Pitts discovered that if individuals who were prone to anxiety were given an infusion of lactate solution, it precipitated an anxiety reaction in some of them. Researchers also looked at other drugs that were known to cause symptoms of anxiety, such as yohimbine, a drug better known as a sexual stimulant in the male, but which caused feelings of anxiety in some individuals. In the early 1980s researchers finally discovered how the benzodiazepines relieved anxiety. It turned out that the anti-anxiety drugs that Sternbach had isolated from a dye compound had receptor sites in the brain. The discovery of the benzodiazepine receptor site led researchers to the conclusion that if there were receptor sites in the brain for benzodiazepines, then the brain must either produce a neurotransmitter that is a benzodiazepine or something very close to it that would fit into the same slot. Further research showed that gamma-aminobutyric acid (GABA), another of the brain's neurotransmitters, worked in conjunction with the benzodiazepines to relieve anxiety. These findings led scientists to another conclusion--- anxiety must have a neurochemical basis. Anxiety is not just *all in our heads*. Scientists also came to a more general conclusion: if there is a chemical that affects the way we feel, that chemical must fit into a receptor site of a neurotransmitter that already exists, or into one that is very similar to it.

About the same time as the discovery of the benzodiazepine receptor sites, another breakthrough in the brain's neurochemicals came about with the discovery of morphine-like compounds in the brain. They were given the name endorphins. Endorphins turned out to be the body's own natural pain killers, and they had their own receptor sites. Both the benzodiazepines and the endorphins are involved in helping the individual adapt to the outside world, one for the relief of anxiety, the other for the relief of pain, and it became apparent during the 1980s that the brain has a neurotransmitter to cope with whatever circumstance the individual might need in order to survive, and stay in homeostatic balance with, not only the internal world, but the external world as well.

The next logical step in the search for the cause of anxiety came about when it was assumed that if there were neurochemicals that relieve anxiety, there must also be neurochemicals that causes us to feel anxious. Feelings of anxiety had to come from somewhere. During the 1980s the old adage that an emotion like anxiety was "all in our heads," was replaced by a new paradigm, all of emotions had to come from neurochemicals that arise from inside the brain. The implications of this were profound, because it meant that the long held belief of a dualism between the mind and body, first espoused by Plato and latter by Descartes, had to be abandoned. Further research showed that all of our feelings, including anxiety, anger, and love have a neurochemical basis. Dr. Candice Pert, who did much of the original research on the endorphins, published *The Molecules of Emotion* in 1997. In her book she makes the case that everything we feel, from anxiety to jealousy, is chemical. Some might find this conclusion disturbing, because we humans like to think of ourselves as being more than just a complex set of neurotransmitters, nonetheless it is becoming more and more evident that this is the case, and even the deepest of our emotions, such as love and hate, have a neurochemical basis.

In the 1970s Dr. Braestrup in Belgium was researching the neurochemical cause for anxiety. He was working with a group of chemicals called beta-carbolines, which had been found in the urine of patients with panic attacks. He found that when he injected volunteers with beta-carbolines they developed intense feelings of anxiety. If they were pretreated with benzodiazepines and then given beta-carbolines, they did not develop the feeling of anxiety. If they were given benzodiazepine after being given beta-carbolines, the feeling of anxiety subsided. Though the question has never been answered as to whether the brain's primary anxiety chemical is a beta-carboline, it is reasonable to assume if the anxiety neurotransmitter is not a beta-carboline, then the brain must have another neurotransmitter or group of neurotransmitters that have the effect of producing what we call anxiety. To carry this line of reasoning one step further, it must also be assumed that the brain not only has a way of turning anxiety on with a anxiety neurotransmitter, but also a way of turning it off through the benzodiazepines and the GABA neurotransmitter system.

As more sophisticated scans were developed to measure brain activity, researchers began looking for areas of the brain that might be involved in anxiety. PET scans, a new type of scanning devices, were especially useful because they were able to measure cell activity by measuring glucose metabolism. This enabled scientist to see which areas of the brain were active during an anxiety attack. One area that showed a great deal of activity during an anxiety attack was in the *pons*, a part of the mid-brain which contains a small bit of gray tissue, called the *locus ceruleus.* Here activity was especially intense during a panic attack. Research also showed that the locus ceruleus has neurons which spread over the entire brain, and many of those neurons transmitted with norepinephrine, a neurotransmitter known to be involved in anxiety. In experimental animals when the locus ceruleus was stimulated, the animals scurried around looking for a place to hide. When this area of the brain was destroyed the animals seemed incapable of reacting to danger. Though there have been other areas of the brain implicated in anxiety, such as the amygdala, the locus ceruleus would seem to be a likely candidate for the primary anxiety center. It is also possible there could be more than one area involved, or a combination of areas; but what is important in developing a theory of anxiety is the realization there is a specific area in the brain that causes an individual to feel anxious through a specific set of neurotransmitters. Once the anxiety neurotransmitters are released from the anxiety center, or anxiety area, what follows is a series of changes that affect the entire body, from the skeletal muscles to the heart, one which prepares the body to get away from harm or danger, or to face it.

Assuming there is such center or area in the brain, let's look at how it should work and then what might go wrong with it. An anxiety reaction would normally be initiated when the brain becomes aware of the possibility that some kind of harm or danger exists. It would usually get this information from the outside through one of the senses, however in some cases it might arise from inside the brain, such as when a thought triggers the feelings of anxiety.

     With the exception of the times when the anxiety center fires off reflexively, as when a loud noise goes off, or a strange face appears in a window, normally the first step after someone becomes aware of the possibility of danger, is for the upper brain, or cerebral cortex, to weigh the data, make a judgment as to the seriousness of the risk, then send a weighted message to the anxiety center to turn on the anxiety neurotransmitters, like turning on a faucet. If the possibility of the injury is slight, a level one, on a scale of one to ten, then the amount of anxiety neurochemical released should be small, a level one release. If the likelihood of injury is great, a level ten, the amount of anxiety chemical released...........

should be very high, a level ten release. In other words, the amount of anxiety chemical released should be proportional to the circumstances, and by definition when it is proportional, the amount of the anxiety felt would be normal. The amount released doesn't have to be an exact match to be called normal; but like the blood sugar, it should fall within a normal range. When the amount doesn't match, either because it's too much or too little, then the amount of anxiety that the individual feels would be considered abnormal.

When the circumstances change, the upper brain's evaluation should also change and send a new message to the anxiety center, causing it to change its output to match the re-evaluation of the upper brain. Experience tells us however that the feeling of anxiety doesn't go away immediately after the circumstances change. It usually takes a while for the anxiety center to calm down and the anxiety chemicals to get out of the system.

Panic Attacks and Post Traumatic Stress Disorder

When the amount of anxiety matches the circumstances, we can say the amount of anxiety is normal, but in individuals with panic attacks and PTSD the amount of anxiety they feel is sometimes off the charts. There is a disconnect between the amount of anxiety they feel and the circumstances. When I have talked to patients about their panic attacks they tell me they recognize the amount of anxiety they have doesn't match the circumstances, but they feel helpless to control it. Some say they feel like they are going crazy. The same thing is true of patients with PTSD. They feel like they are in great danger and are going to die for no apparent reason. The number of anxiety neurotransmitters in their brains must be very high. But why? It doesn't make sense for the brain and body to be flooded with anxiety neurotransmitters when the situation doesn't call for it. Something in their anxiety system goes haywire. But what is it?

There can be two, and only two, processes involved in the production of anxiety. One is the evaluation of the circumstances by the upper brain, or cognition, and the other is the output of anxiety neurotransmitters by the anxiety center; and if cognition is not the problem, and someone has too much anxiety, the culprit must lie in the anxiety center itself. This means that the only plausible explanation for panic attacks and the relentless anxiety that patients with PTSD feel, is overproduction of anxiety neurotransmitters by the anxiety center. Cognitive errors are not the problem in either panic attacks or PTSD. In panic attacks the anxiety center overreacts to a stimulus from the upper brain and puts out far too much anxiety neurotransmitter for the circumstances. In PTSD the anxiety center not only overreacts to stimuli, but in some individuals it is so unstable that it sometimes fires off without any stimuli at all, causing them to feel anxious for no apparent reason.

Someday in the future, after an anxiety neurotransmitter has been identified, I think it will be possible to accurately measure the amount of anxiety that someone has by drawing a blood sample and measuring it in the lab, but we aren't there yet. The best I have been able to come up with is a simple one-to-ten scale that can be used to measure both the importance of an event by the upper brain, and to estimate the amount of anxiety neurotransmitter released from the anxiety center. In practice the one-to-ten scale works pretty well.

A New General Theory of the Primary Survival Emotions

My hypothesis is that there are two parts of the puzzle that explain why emotional responses sometimes don't match the circumstance, and by definition are abnormal. The first part of the puzzle is explained by errors in cognition, the weighing process which occurs in the upper brain, or cerebral cortex. When cognition is faulty an incorrect message will be sent to one or more of the primary survival emotional centers, which will cause an error in the type and/or the amount of the emotion released. This is a cognitive error, and the type of problem that Drs. Ellis and Beck and others address with rational emotion therapy, or cognitive behavioral therapy. When therapy is successful an individual's cognition is changed to reflect a more realistic and less distorted view of the world, and subsequent messages sent to the emotional centers will be more accurate, and their emotions will more closely match the circumstances.

The other part of the puzzle, which explains why the amount or type of an emotion sometimes fails to match the circumstance, has nothing to do with cognition, but instead involves the primary survival emotional centers themselves. Even though one of the primary survival emotional centers gets the correct message from the upper brain, it may not work correctly, and may produce too much of the neurotransmitter that gives rise to a particular emotion, which is what happens in a panic attack when the anxiety center floods the system with anxiety neurotransmitters after a stimuli.

A third cause for abnormal emotions is the result of instability of one of the primary survival emotional centers which causes it to produce its emotional neurotransmitter spontaneously without any stimulation from the upper brain. One example is the instability of the manic and depression centers that causes the spontaneous mood swings that characterizes bipolar disease. This theory also offers an explanation for the mixed state, which occurs when someone with bipolar disease feels both manic and depressed at the same time. At present there is not a good explanation for the mixed state, but it can be easily explained if we assume that sometimes both centers are active at the same time. Another example of emotional center instability is in major depression, or unipolar depression, which is caused by a malfunctioning depression center that spontaneously pumps out depression neurotransmitters over a lifetime. Major depression is an ailment that has nothing to do with cognition, and this is the reason why psychotherapy doesn't work in major depression. Not even the most intensive psychotherapy, or psychoanalysis, can turn off a malfunctioning depression center.

One of the aims of this project is to move psychiatry and psychology from the realm of the qualitative into the realm of the quantitative. Until now psychology and psychiatry have been more metaphysics than physics, more ideas than substance, and even though the type and amount of an emotion can't be measured in the lab as yet, I believe that one day we will be able to measure both and when that day comes psychology and psychiatry will move even closer to the realm of physics, but for now the one-to-ten scale will have to do.

In summary:

(1)   I propose there are *six primary emotions*---four survival emotions---anxiety, anger, depression and manic, all of which were built in and have specific centers in the brain; and two non-survival emotions--- happiness and sadness,

and through education and inquiry it is possible to identify each emotion with certainty, and they not be subject to uncertainty and interpretation as is the case today.

(2)   I propose that the *amount* of a particular emotion can also be measured, and even though we can't get an exact measurement from the lab today, we can usually get a pretty good idea by using a one-to-ten scale.

(3)   I propose the *significance, weight, or importance* that someone places on an event, or circumstance, through the process of cognition, can also be measured and weighed.

(4)   I propose that a primary emotion is *normal* when the amount and type of the emotion is proportional to the circumstances and falls within a normal range.

(5)   I propose there are four causes for an abnormal emotion:

(a)   The first is due to an error in cognition and occurs when the upper brain fails to weigh circumstances correctly and sends an incorrect message to the right emotional center, but this causes the center to put out an incorrect amount of the right emotion, usually too much. This is what happens in social anxiety where the individual places too much importance on being liked by everybody, or has the unreasonable fear that they will commit some social *faux pas*. These are cognitive errors.

(b)   A second cause is also a cognitive error and occurs when the upper brain fails to weigh circumstances correctly, but this time sends the message to the wrong emotional center, which causes it to put out the wrong emotion. This is what happens when someone loses a love one to a heart attack, or an unavoidable accident, but the individual feels they have to blame something or someone, it has to be somebody's fault. This kind of thinking leads to feelings of anger, rather than the emotion which would be normal, sadness.

(c) A third cause is when the upper brain weighs the circumstances correctly, and sends the correct message to the right emotional center, but the emotional center malfunctions and overproduces its emotional neurotransmitter. This is what causes a panic attack. Cognition is not the problem in panic attacks.

(d) The final cause is when one of the primary survival emotional centers is unstable and discharges its emotional neurotransmitter spontaneously. This is what causes the anxiety and depression in bipolar disease, major depression, and PTSD. Cognition is not involved in any of these problems. They are all due to emotional center malfunction.

With the exception of anger-aggression, determining a normal level of the other primary survival emotion is fairly straight-forward; however, with anger-aggression the issue is more problematic, but I will address this in more detail in the section on anger-aggression. Using the medical model, when one of the body's functions falls outside the normal range we call it a disease, and I see no reason not to use the same terminology when the amount or the type of an emotion falls outside the normal range. This is in opposition to the DSM which dodges the issue and labels them disorders. When one of the glands of the body malfunctions and puts out too much or too little of its hormone, or neurotransmitter, we call it a disease, a state of dis-ease. When the thyroid gland overproduces we call it Grave's disease, we don't call it Grave's disorder. So in order to make things consistent I believe we should call mental or emotional illnesses diseases, just like we do the physical ones.

Making a Distinction between Mental and Emotional Illness

When we talk about ailments like depression and anxiety, sometimes we use the term mental illness, sometimes emotional illness. The two terms are used indiscriminately, but in order to clarify the way we use and think about these terms, I would suggest that the term emotional illness be used in those illnesses where there is a malfunction of one or more of the primary survival emotional centers, and the cognitive or mental processes of the upper brain are not involved. A good example of a pure emotional illness is bipolar disease, where the manic and depression centers are abnormal. Other examples of emotional illnesses are PTSD and panic attacks, where the anxiety center is abnormal. I would restrict the term mental illness to those illnesses where errors in cognition are involved. Examples would include generalized anxiety disorder and some cases of depression. When someone has elements of both a mental and emotional illness, I suggest we use the term *mental-emotional illness*. For example, not infrequently someone with an anxiety disorder will have both mental and emotional aspects to their illness, such as when someone inherits an overactive anxiety center from a mother who also has an overactive anxiety center, and has been programmed by the mother to see the world as very scary place, leading to cognitive errors

Bipolar Disease as an Example of Emotional Center Malfunction

A classic example of emotional center malfunction is bipolar disease, which is caused by instability of the manic and depression centers. The original term for this condition was manic depressive disease, and descriptions of it go back to antiquity. Hippocrates wrote about it more than 2000 years ago. The term bipolar disease, or bipolar disorder, came from a paper written in 1953 by Dr. Karl Kliest, a psychiatrist, who suggested that manic depressive disease was like a person with a manic pole and a depressive pole. If someone had a problem with only one pole he called it unipolar disease, but when both poles were involved, he called it bipolar disease. When the DSM III was updated and republished in the 1970s, Dr. Kliest's ideas were incorporated into its nosology, or classification, and the term manic depressive disorder was dropped and replaced by the term bipolar disorder.

Dr. Kliest was not that far off and would have been correct if, instead of two poles, he had proposed two distinct emotional centers. His theory of two poles was important in explaining how someone's emotions can shift back and forth, but his theory fails to explain the mixed state, which is when someone with bipolar disease feels both manic and depressed at the same time. The mixed state can be explained if we assume that instead of there being two poles there are two distinct emotional centers, and the mixed state occurs when both centers are active at the same time.

It is important for someone with bipolar disease to realize they have a condition caused by the malfunction of two of their survival emotional centers and has nothing to do with how they think, or their cognitive functioning. It is therefore not surprising that psychotherapy in bipolar disease is worthless, other than helping the patient understand that there is something wrong with two of their emotional centers. Freud discovered early on that psychoanalysis was not helpful in patients with this illness, and he quit trying to treat them. Today we know that the treatment of bipolar disease is medication, and fortunately we have at our disposal a number of medications that can treat it very well.

Tripolar and Quadripolar Disease

Anger and irritability are listed in the DSM as one of the symptoms of bipolar disorder, but here the DSM makes another mistake when it states that anger and irritability are a *symptom* of bipolar disease. When someone with bipolar disease has spontaneous periods of anger and irritability, it is not because their anger and irritability are symptoms of their bipolar disease; it is because they have an anger-aggression center that is also unstable, just like their manic and depression centers.

Carrying this line of reasoning one step further, I propose another hypothesis about the primary survival emotional centers. In addition to there being what is currently called bi-polar disorder, which is caused by the instability of two of the primary survival emotional centers, some individuals have instability of three of their primary survival emotional centers, resulting in what we could call tri-polar disorder. There are also some individuals who have instability of all four of their primary survival emotional centers, resulting in quadripolar disease or disorder.

I am currently treating a young man who came to the office originally for help with his anxiety. He is a jailer and works with some of the worse inmates in a lock up unit. At times when he was at work he would suddenly feel very frightened and develop the classical symptoms of a panic attack. I treated him initially for the panic attacks, but then other symptoms began to come out. It turns out he also had periods when he felt very depressed for no reason, and other periods when he would feel overly confident and full of energy, symptoms that were consistent with bipolar disease. But he also had times when he would become very angry and develop attacks of rage that seemed almost out of control. These would occur mostly at work, but sometimes at home with his wife and kids. It was apparent this patient had instability of all four of his primary survival emotional centers, resulting in quadripolar disease. In his case the anxiety, depression and manic symptoms have been brought under control with medication, but the control of his anger has proved to be more difficult, and I am still working with him to try and find a medication that will help his anger.

I have become increasingly aware of how many patients have instability of more than one of their emotional centers. In fact when I see someone with one emotional center that is unstable, more than likely they have others as well

Anxiety Center Malfunction

If we assume that my theory is correct, and that abnormal primary emotions can occur from either emotional center malfunction or cognitive errors, how might this play out in the anxiety disorders listed in the DSM? Though it is theoretically possible for the anxiety center to produce too little anxiety neurotransmitter, most of the time when the anxiety center malfunctions it puts out too much and causes the individual to feel too much anxiety, as in a panic attack.

The DSM lists the following types of Anxiety Disorders:

(1)   *Panic disorder without Agoraphobia*

(2)   *Panic disorder with Agoraphobia*

(3)   *Agoraphobia without a history of Panic Disorder*

(4)   *Specific Phobia*

(5)   *Social Phobia*

(6)   *Obsessive-Compulsive Disorder*

(7)   Posttraumatic Stress Disorder

(8)   Acute Stress Disorder

(9)   Generalized Anxiety Disorder

(10)   Anxiety Disorder due to a[Medical Condition or Substance Abuse Related

(11)   Anxiety Disorder not otherwise specified.

Let's take each of these diagnoses and see whether the problem might relate to a malfunction of the anxiety center, or a cognitive error, or both.

(1)   A panic attack without agoraphobia occurs when the anxiety center suddenly releases a large amount of anxiety neurotransmitter when the situation does not call for it, and a crowd is not involved. Cognition is not the problem.

(2)   A panic attack with agoraphobia occurs when the anxiety center releases a large amount of anxiety neurotransmitter in the setting of a crowd or market place. *Agora* means market place in Greek, today it would be the mall. This also is the result of anxiety center malfunction and has nothing to do with cognition.

(3)   Agoraphobia without a history of panic is a cognitive problem, and occurs when someone sees the mall as a very scary place, knowing if they go there their anxiety center will likely fire off.

(4)   Anxiety with a specific phobia occurs when the anxiety center fires off when exposed to a specific trigger such as heights, spiders, snakes, etc., and where there is no real danger to the individual.

(5)   Social phobia involves a combination of anxiety center malfunction, and a cognitive disorder. It occurs when an individual places too much emphasis on being liked by everyone, or are overly concerned they will make a social *faux pas* of some kind, but who also has a jumpy anxiety center. This is both a cognitive problem and an anxiety center problem.

(6)   Obsessive-compulsive disorder is really two problems. The compulsive aspect is an illustration of *magical thinking*. Individuals who go through rituals like washing their hands

multiple times, or checking behind themselves numerous times to make sure that they have, or have not done something, do so in an effort to try and magically keep their anxiety center from firing off. They believe that as long as they continue the ritual it will keep their anxiety center from firing off.

The obsessive aspect of this illness is a different problem altogether and has something to do with some kind of reverberating circuit in the brain which causes a thought to go around and around in someone's head which they can't stop. This eventually leads to their anxiety center firing off when they can't stop the recurrent thought.

(7) Post traumatic stress disorder is the result of a broken anxiety center. Initially an individual is exposed to a traumatic event of high intensity which causes the anxiety center to produce a very high level of anxiety neurotransmitter, which matches the circumstance and is appropriate at the time; but this event damages the anxiety center and causes it to becomes unstable and start firing off at the slightest provocation, and at inappropriate times.

(8) Acute stress reaction is when an individual is in a very stressful situation and should have some anxiety, but has more than the situation calls for. This could result from either a cognitive problem, or an emotional center malfunction.

(9) When someone is sick it is normal to have some anxiety, but determining how much anxiety is normal means measuring the amount of anxiety they have against the seriousness of the medical condition and seeing if the two match up. For example, if someone is told that they have cancer, a high level of anxiety would be normal, say a nine or ten. But if someone is told that they have strep throat, having a level ten of anxiety would not be normal. But the circumstances must always be considered in the equation. If the person with the strep throat is a young opera singer, and his or her big chance to fill in for the star the next day is dependent on them being able to perform, then a high level of anxiety under such circumstances would be normal for them. Anxiety which is associated with substance abuse is more than likely a result of stimulation of the anxiety center by whatever drug is involved. Cocaine, for example, causes

some individuals to feel very anxious probably through its stimulation of the anxiety center.

(10) Anxiety Disorder Not Otherwise Specified is a catch-all term for cases of anxiety that don't fall into any of the other categories.

Some Additional Thoughts on Agoraphobia and Post Traumatic Stress Disorder

The term agoraphobia comes from two Greek words agora, which means the market place, and phobia which means the fear of something. Today it refers to the fear of any kind of crowd. Patients with agoraphobia don't like to go where there are a lot of people. I have had patients with agoraphobia tell me that they could go to the grocery store, walk up and down the aisles without any trouble, but when they got in the checkout line and were surrounded by people they would develop a panic attack and have to rush out of the store leaving their groceries behind. In the worst cases of agoraphobia, the individual becomes so paralyzed by their own anxiety center that they are unable to leave the house, and may even be confined to a single chair in the house, because even leaving the chair will cause them to have a panic attack. When someone has agoraphobia to this degree it is because they have an anxiety center that is so sensitive that the moment they get out of the chair their anxiety center fires off. That chair is their only safe place, and the only place where their anxiety center is quiet.

Agoraphobia of this severity is not that uncommon, but unfortunately most people who have it this degree are reluctant to come to the doctor and get treated. They often spend their lives confined to their homes, terrorized by the fear of their own anxiety center and convinced that they can't be helped. Howard Hughes probably had agoraphobia to this degree. Individuals with agoraphobia can be helped, sometimes through desensitization, sometimes through medication, and sometimes through both. It is interesting to note that many of the drugs, such as the benzodiazepines and the barbiturates, which will treat anxiety, are also the same drugs we use to treat seizures. Panic attacks are probably caused by some type of seizure in the anxiety center and we don't hesitate to use drugs to control seizures, so why should we be any less hesitate to use drugs to control an overactive anxiety center. There are some individuals, both professional and lay persons, who feel that individuals who have panic attacks should just tough it out and not take medication, not realizing that these individuals are living in a kind of private hell over which they have no control, and who could be helped with medication.

When I see a patient with agoraphobia and panic attacks the approach I take with them is the same I take with a patient with diabetes. I explain to them that the anxiety they feel is not something they are imagining, but it is because they *have* too many anxiety neurotransmitters in their system, the same way that a patient with diabetes *has* too much blood sugar. And the reason for this is because their anxiety center is broken, not unlike the way a diabetic's pancreas is broken. To help them understand what is happening I compare their anxiety center to a dimmer switch. As the demand for anxiety goes up, the anxiety center should increase the number of anxiety neurotransmitters the way a dimmer switch increases the amount of light. But in patients with panic attacks and agoraphobia, their anxiety center has lost its ability to function as a dimmer switch and has begun to function as an on-off switch. It is either all the way on, or all the way off. Patients often nod their heads in agreement when I explain it this way. They tell me that this is exactly what it feels like when they have a panic attack.

PTSD is another example of a broken anxiety center. In PTSD the individual is initially exposed to some horrific event, one that causes the anxiety center to put out a huge amount of anxiety chemical, an amount that at the time is appropriate, but the shock to the anxiety center somehow breaks it, causing it to become unstable and too sensitive. It starts over producing anxiety neurotransmitters and causing panic attacks. Sometimes it fires off at inappropriate times, putting out anxiety chemical for no apparent reason. It can also lose its ability to function like a dimmer switch and begin to function as a very unstable on-off switch. It will either be all the way on, or all the way off, with no in between. Interestingly this same phenomenon can happen to a real dimmer switch when it is overloaded with voltage. It can lose its ability to function as a dimmer switch and begin to function as an on-off switch.

The Neurotransmitters of the Primary Survival Emotional Centers

Today the research on the neurotransmitters of the emotions is focused primarily on serotonin, dopamine, norepinephrine, glutamate, GABA, and some others. Emotional diseases such as depression, bipolar disease, and panic attacks have been attributed to *chemical imbalances*, but this explanation is overly simplistic and doesn't really explain what is going on. Current research has been successful in identifying many of the trees of the mental-emotional forest, but the forest itself remains obscure. In the depression part of the forest we know that individual with depression have elevated cortisol levels. (Interestingly, cortisol levels are also elevated in hibernating bears, which adds to my suspicion that there is a connection between depression and hibernation.) We know there are other changes in the brain in patients with depression. The hippocampus, for example, is smaller in patients with depression, which may be secondary to decreased level of BDNF, which stands for brain-derived neurotrophic factor. For years we were taught that once we reach adulthood, we quit producing new brain cells, but it turns out that this assumption was wrong. Research by Dr. Fred "Rusty" Robert Gage has shown that we continue producing new brain cells as long as we live. It is BDNF that promotes new brain cell growth and in depression BDNF levels are down, and this probably explains the atrophy of the hippocampus that is seen in patients with depression. When patient with depression are treated with antidepressants, BDNF levels rise, and the atrophy of the hippocampus reverses. Some have speculated that decreased levels of BDNF might be one of the causes of depression, but I suspect that it is the other way around, and that a depression factor suppresses BDNF. If a depression-hibernation factor was built in to help us survive helpless, hopeless circumstances by slowing everything down, then it would make sense that when we were hibernating we wouldn't need new brain cells and BDNF would be suppressed.

Changes are seen in the MRI and PET scans of patients with depression, but the underlying causes for these changes remain obscure, and articles on depression readily admit that the underlying cause of depression is unknown. In Atlanta, Dr. Helen Mayberg, professor of psychiatry and neurology at Emory, has been working with an area of the brain call Cg25, and its role in depression. Pet scans have shown that this area is especially active in patients with depression. Using the same technique that is sometimes used on patient with Parkinsonism, she and some of her colleagues tried deep brain stimulation of area Cg25 on six patients with treatment resistance depression to see what might happen. When this area was stimulated four of the six patients said that it was as if a switch had been turned off, and the depression they had felt for years just disappeared. This kind of relief had never been seen before with any kind of treatment, even ECT. In her comments on this unique treatment, Dr. Mayberg said that she couldn't be certain whether deep brain stimulation of area Cg25 was causing an inhibitory effect, or a stimulating effect, whether it was turning something on, or turning something off, and said it would take further research to tell exactly what was happening. I speculate that the stimulation of area Cg25 was turning off a depression factor or the effects of some kind of depression factor.

The closest analogy to depression in humans is hibernation in animals, and research has shown that there is a specific hibernation factor in the brains of hibernating animals which causes them to hibernate. If such a neurotransmitter exists in the brains of hibernating animals, it is reasonable to assume that something similar exits in the human brain. This would fit in with a theory of the emotions that posits that each of the primary survival emotions was built into the brain to serve an evolutionary purpose as Darwin proposed. Where depression is concerned, there must have been a time in our evolution when our ancestors needed such a center to slow things down in the wintertime, or other harsh conditions, in order to conserve energy and enable our ancestors to survive.

The Master Emotional Neurotransmitters

I propose that each of the primary survival emotions has a specific center in the brain that initiates an emotion by releasing a master emotional neurotransmitter that is unique to that emotion, and what follows is a chain of commands involving many different neurotransmitters that carry the emotional message from the brain to the different parts of the body. Like the chain of commands in the Army, the anxiety general would give a command to the colonels, which might be serotonin, and on to the majors, which might be norepinephrine, and so on down the line to the individual muscles or organs. Some of the same intermediate neurotransmitters must be involved in more than one of the primary survival emotions. Serotonin, for example, must be involved in both anxiety and depression, because we know that the SSRIs can treat both anxiety and depression. A single neurotransmitter type, such as norepinephrine, also must carry the message to multiple sites in the body, because we know norepinephrine affects both the heart and blood vessels.

When we feel one of the primary survival emotions we are feeling a specific blend of neurotransmitters coursing through our brains and bodies that is distinctive to that emotion. When we touch a hot stove, and feel heat and pain, it is because the nerve endings in our fingers bring a message to the brain from the heat and pain sensors in the finger tips. When we feel an emotion, it is because our brains and bodies feel the neurotransmitters of that emotion, and the effects that those emotional neurotransmitters have on our brains and bodies. When we feel anxious, we feel the pounding of the heart in the chest, and tension in the muscles from epinephrine and norepinephrine, all part of the anxiety complex. When we feel anything, whether it is an emotion, or the burn from a hot stove, it is because the brain is registering a particular neurotransmitter state.

In Dr. Antonio Damasio's book *Looking for Spinoza* he writes about how the brain feels, and makes a distinction between how the brain registers feelings and how the brain registers emotions. Dr.

Damasio places an emphasis on the interpretation by the brain of the different feeling states and for him it is the interpretation of the emotional neurotransmitters that is important, not the neurotransmitters themselves. But this is basically a reworking of the James-Lange theory put forth over a hundred years ago, when James said, *I feel sad because I weep, not I weep because I feel sad,* and as I pointed out earlier the James-Lange theory falls apart when it tries to explain why we sometimes weep out of joy. When we weep for joy, is it because we are really sad, but just don't recognize the sadness? I don't think so.

It is not uncommon for someone's upper brain, I, cerebral cortex, ego, conscious self, or however it is labeled, to fail to indentify a particular emotional neurotransmitter state. Someone can label anxiety as depression, and anger as anxiety. My thesis is as follows; just because someone misinterprets a particular emotional neurotransmitter state, does not change the emotion, the emotion remains the same. I often have to help patients understand which emotion they are actually feeling, because they often confuse one emotion for another. One of the objectives of this book is to help educate the reader to a better understanding of the emotions and to help them be able to identify which emotion they are actually feeling.

Dr. Damasio is correct about one thing, and that is Spinoza was right and Descartes was wrong. Descartes felt that the mind and the body were two separate identities; a concept first expressed by Plato, but today indentified more with Descartes. Descartes felt that the mind was a separate entity from the body, and could exist apart from the body. Spinoza, a contemporary of Descartes, believed that the mind and the body were part and parcel of the same thing. Today we know that the mind cannot exist apart from the body, because the mind is a product of the brain, which is a part of the body. When the brain deteriorates, so does the mind. The classical example of this is Alzheimer's disease, a deteriorating brain disease of unknown cause, which slowly destroys the individual's brain, and with it, his or her mind.

Anger-Aggression, the Emotion of Possession and Control

Anger-aggression is the one human emotion that has caused mankind more grief and misery than all of the earth's natural disasters put together. Since time immemorial mankind has always had trouble keeping his anger and aggression in check. It has been the cause of history's wars, rapes, murders, genocide, conquest, bullying, abuse, corporate greed, torture, water boarding, abuse of women and children, invasion of other countries, and man's inhumanity to nature. Society has always had a great deal of difficulty in deciding just how much anger-aggression to allow, and almost all laws are written in an effort to set limits on man's aggressiveness, from the setting of speed limits, to the prohibition against murder.

Some of the words and phrases we use to describe the range and the effects of this emotion are: to blow one's top, fume, go berserk, lawlessness, throw a fit, abuse, evil, rant and rave, scold, wrathful, irritable, petulant, anger, acrimonious, kill, barbaric, mean, rape, murder, martial, hawkish, disruptive, attacking, contentious, rancorous, sarcastic, jealous, pushy, advancing, pugnacious, hostility, antagonistic, assertive, aggressive, offensive, belligerent, ill, intrusive, hostility, combative, destructive, fight, mad, spiteful, greed, pushy, sadistic, angry, rage, pissed off, punish, pushy, genocide, and so on.

All of these words express the same basic emotion and only differ in the amount they represent. With frustration and annoyance, there is a small amount, with anger and rage much more. Just as society has had a great deal of difficulty in knowing what to do with this emotion, so too has psychiatry. In psychiatry anger and aggression are treated as a *symptom* of one of the other emotions, but never as a separate emotion. For some reason or other this emotion has never been given the same status as some of the other emotions, like anxiety and depression. This apparently is because psychiatry doesn't quite know what to do with it either. But, by relegating it to the status of being only a symptom of one of the other emotions, its importance has been greatly underestimated. When anger accompanies depression, it is important that the anger-aggression be dealt with as a separate emotion, and not as a symptom of the depression. Patients with bipolar disease often have a great deal of anger, and when they do, there are three emotions involved— depression, manic, and anger. If the anger is lumped in with the other two emotions, and only two emotions identified, it diminishes the importance of the anger. The DSM makes this mistake when it list irritability and anger as a symptom of depression. This is especially true when it discusses depression in teenagers which is a terrible mistake.

When we feel anger or aggression, frustration or annoyance, it is because we are feeling different amounts of an emotional neurotransmitter which originates from a center located somewhere in the lower brain. This center was built into our brains millions of years ago as a mechanism to help us survive, and like anxiety and depression its effects are mediated by a master anger-aggression neurotransmitter which gives rise to a cascading series of events that causes us to feel this emotion. Without this emotion and its physical effects, our ancestors wouldn't have survived very long. They needed a certain amount of anger-aggression in order to obtain food and shelter, protect family, and maintain territory. We still need a certain amount of aggressiveness, or assertiveness in order to survive in today's world, but unfortunately all too often the amount of aggression and anger seen in today's world is excessive. Throughout history mankind has shown himself to be a very aggressive creature, and it is no less true today than it was six thousand years ago when man first started recording history. One has to look no farther than the morning paper, or watch the news on television, to see just how true this is. The vast majority of the news is devoted to examples of excessive aggression and anger in one form or another. There are stories of war, rape, murder, company greed, cheating, stealing, bullying, torture, suicide bombings, genocide, and on and on; all forms of aggression of one form or another.

When we feel any emotion, be it anger, depression, or anxiety, it is because there is a specific blend of neurotransmitters in our brains and bodies that is specific to that emotion; and it is especially important where anger is concerned to note there is a difference between having the neurotransmitters of anger in our brains and bodies, and acting on those neurotransmitters. Just because I feel angry doesn't mean that I have to act *angrily*. If I feel anger toward someone, it doesn't mean that I have to hit them, or worse, shoot them. The same is true of anxiety. Even though I might feel anxious doesn't mean that I have to let my anxiety keep me from getting up in class to recite, or making a presentation at the office. It is well to remember that when we feel any emotion it is wise to let the upper brain evaluate the circumstances before acting on the emotion. This is especially true when the emotion is anger. The prisons are full of people who have acted on their anger neurotransmitters without regard to the consequences.

I have found it helpful in developing a better understanding of the emotions to look at the derivation of the word from which a particular emotion was derived. The meaning of words often changes over time, and not everybody understands the same thing when we use a particular word. It interesting to note that anger and anxiety have their roots in the same word. The word anger arose in the German and Scandinavian languages around the 13th century from the word *angre* or *engi,* whose original meaning was to feel grief or distress. Over time it evolved into its modern meaning, which is not so much to feel grief, as it is to cause grief or distress. The word assertive comes from the Latin word *assertus*, which means to lay claim to. Assertive was first used in the English language in the 1560's. The word aggression comes to us from Latin through the French. *Aggressus* in Latin means to approach or move forward and was first used in the English language in the early 1600's.

Even though at present we can't measure how much anger someone has in the lab, we can get a pretty good idea by observing the way someone looks and acts. When someone is really angry, the blood vessels in their face will dilate and their face turn red, their muscles will get tense, their eyes will narrow, and their voice will change. From observations like these we can get a rough idea of how much anger a person has, but we need to remember that an individual's anger can be hidden or suppressed, so it is often difficult to tell just how much anger a person really has, especially if they want to hide it.

When we observe the world around us we see aggressive behavior everywhere. From early childhood most boys are taught to be aggressive, and it would appear that in today's world girls are becoming more and more aggressive as well, hence the term *mean girls*. A certain amount of aggressiveness or assertiveness is admired by society and this is especially true in athletics, and with the exception of a few teams like the Maryland Terrapins, and the Saint Louis Cardinals, I can't think of many athletic teams that don't have very aggressive names and mascots. There are literally thousands of teams from high schools to the pros with names like Bulldogs, Timberwolves, Copperheads, Cobras, Panthers, Lions, Hawks, Eagles, Stallions, and in general the more aggression the name represents the better. It would be hard to imagine the Chicago Cows. Our local hockey team is named for one of the most aggressive and destructive forces in nature, the Hurricanes.

But how much aggression is normal, and when does it cross the line? One of the objectives of this project has been to develop a theory of *normal* that can be applied to all of the emotions, and one of the criteria that I have used to define normal is to say that an emotion is normal when the type of the emotion, and the amount of the emotion matches and is proportional to the circumstance. For most of the primary survival emotions saying how much of a given emotion is normal is fairly easy, but in the case of anger-aggression the answer is more problematic. Society has many laws that establish guidelines for how much aggression is permitted, but there is not a general theory of how much aggression is normal, or when it is abnormal. If you are driving along, and another driver pulls in front of you, just misses you by inches, and runs you off the road; some degree of anger, along with some anxiety, would certainly be normal, but if someone pulls out a gun and shoots the driver of the other car that would be an example of excessive anger. Unfortunately this has happened all too often in real life instances of road rage.

Assertiveness, a medium level of this emotion, is admired in politics and the business world, and a certain degree of assertiveness is necessary to get ahead in this world, but when the level increases too much, it can lead to what happened at Enron and Wall Street, when some individuals became overly aggressive, and greed drove some of them to outright fraud. If someone tries to harm one of our children, it is normal for us to have very high levels of anger, but most situations would not call for us to actually kill the person, but that too has happened.

Evil and Anger-Aggression

As I was contemplating the problem of aggression, it occurred to me that in every act of evil there is always an element of aggression, either active or passive. I couldn't think of a single example of evil where aggression was not involved in one form or another, and it became apparent that *all* examples of evil could be defined in terms of aggression. I remembered reading an article about evil in the New York Times magazine section a number of years ago. It was written soon after the Jeffrey Dahmer murders, and about the same time that a mother in South Carolina drove her two children into a lake, sat on the bank, and watched them drown. The author wrote that he didn't believe that evil could be explained, and went on to say that evil was too mysterious, and speculated maybe there is a Satan who makes people do evil things.

The word evil comes to us from the Old English. It first appeared in the English language in the eighth century and is mentioned in the early English novel Beowulf. The original word was spelled *yfel*, and its original meaning was bad, wicked, and vicious; and even today the word carries with a certain mystery and dread that most people associate with Satan. But the more I thought about evil and aggression, the more I realized maybe it wasn't so mysterious after all if we defined it in terms of anger and aggression.

The one element that must *always* be present in every act of evil is aggression in one form or another; and without it no act of evil can exist, therefore evil *must* be a form of aggression. The second element that must be present in every act of evil is a differential in the power base between the evil doer and the victim, and as the power differential widens, the potential for evil goes up. The third thing element that must be present in every act of evil is a deliberate intent on the part of the evil doer to harm the victim, either directly, by some aggressive act on the part of the evil doer; or indirectly, by some inaction on the part of the evil doer that leads to the victim being harmed by some external aggressive force that the evildoer could prevent if they chose to do so. I also came to the conclusion that evil is not an *either/or* thing, as some have suggested, but that evil exists in degrees and can be measured from a little bit of evil, to a great deal of evil. It is also not the case that someone *is* evil, or *is not* evil. Evil is something that someone *does,* not *is.* I would offer the following mathematical formula to measure the amount of evil in a given act.

$$|E| = [(Ped \times Aed) - (Pv \times Av)] \times Hv$$

Where:

$|E|$ = the total amount of evil in a given act

Ped = power of the evil doer

Aed = the amount of aggressiveness or force that the evil doer exerts in an effort to harm the victim

Pv = power of the victim

Av = the amount of force that the victim is able to resist the evil doer

Hv = intentional harm done to the victim

It is apparent from this formula that the greater the power of the evil doer, and the lesser the power of the victim, the greater the potential for evil. One example of great evil is when an adult deliberately harms an infant, because the adult has all of the power and the infant has none, and therefore is unable to resist. At the other end of the spectrum, if the evil doer and the victim are equal in power, or close, then the amount of evil that might result from a confrontation might be negligible, even though there might be a great deal of aggression involved in the confrontation, and the intent to harm is present. But when they are equal in power, the equality in power cancels out the evil. For example, if two prize fighters with equal skills and power come together in the ring and beat the hell out of one another, the intent of harm is there, but evil would not be involved because a power differential between the two did not exist. But if we change the formula and have one of the prize fighters arrange for someone to put knock out drops in the water of his opponent, this changes the power differential and now when they fight his opponent will be weaker and unable to protect himself, and evil will be involved. Something like this actually happened several years ago in a skating competition when the manager of one of the competitors actually took a baseball bat to the knees of his skater's rival in an attempt to weaker her so that his skater could win. This was an evil act from several points of view.

There is one form of evil that involves a different kind of aggression on the part of the evil doer. Passive aggressive behavior can be just as evil as active aggression can be. For example, a man is walking along a sidewalk and sees a baby carriage with an infant in it rolling down a drive way which leads into a busy thoroughfare. He sees the mother running frantically to catch the carriage, but sees that she can't catch it before it rolls into the street. If he stands idly by and watches the carriage roll into the street and be smashed by a car, he hasn't done anything, but he has still committed a very evil act by not stopping the carriage, because he has the power to do so. The aggressive force in this case is the force of gravity, but gravity is not the evil doer; it is the one who has the power to prevent harm from coming to the infant, and instead lets it happen.

Sometimes the power deferential between a victim and the evil doer will shift back and forth. The victim becomes the evil doer, and the evil doer becomes the victim. This scenario has been depicted in films a number of times. I remember one film in which a rape victim turns the tables on the man who raped her. She manages to get the upper hand, and when she does, she now has all the power and he becomes the victim of her evil acts.

Writers for the movies and TV are always looking for new examples of evil doers, and new variations on evil. Their evil doers have included worms, birds, snakes, aliens, tornadoes, volcanoes, earthquakes, serial killers, children, even babies. Another twist on evil, which often occurs in the movies, is the plot in which someone first appears to be innocent, but turns out to be an evil-doer in disguise. There was a movie several years ago in which a young boy turns out be Satan, or the son of Satan in disguise. I don't remember which, I didn't see the movie, I only remember reading about it.

One of the requirements of an evil act is a deliberate intent to harm the victim. If the intent to harm is absent, evil is not involved. Not too long ago a heart transplant was botched at a local medical center because the patient who was a young girl got a heart of the wrong blood type. The surgeons had all of the power and the little girl had none. Though they acted very aggressively toward her in an effort to replace her defective heart, she unfortunately got the heart of the wrong blood type and died from a transplant reaction. She was harmed, but evil was not involved in this case because there was no intent to harm. Negligence was involved in the case, because someone forgot to check the blood type, but not evil. Sometimes the difference is subtle.

There are a number of reasons why individuals commit evil acts. Sometimes it is for the feeling of power and gain they get from being able to harm their victim. Addiction disease is often involved in some cases of evil. In serial killers the motivation to kill and maim others is probably due to an addiction to the feeling of power and evil. Jeffrey Dahmer undoubtedly had an addiction to the feeling that he got when he killed his victims and did terrible things to them. In his case he began torturing animals when he was a boy, and when he got older his addiction worsened, and he turned from animals to people.

Bullying is another example of evil. It often begins in childhood and often carries over into adulthood. Studies have shown that boys who are bullies in grammar school often grow up to be wife beaters as adults. Bullying among school children is one form of evil that has been tolerated for far too long. Today there are programs in schools which address this issue and teach children how not to bully, or how to stand up to bullying. Hazing and some forms of initiations are also acts of bullying and should no longer be tolerated.

Churches sometimes take the view that evil is something that enters someone from the outside like an infection, just as mental illness was once thought of as being caused by something that enters the individual from the outside. Today we know that most mental illness is not caused by something that enters someone from the outside, but instead comes from something that goes wrong inside the brain. Evil doesn't comes from the outside either. The Devil, or Satan, doesn't make people *do it*. Human evil is always the result of someone intentionally harming someone else who is weaker than they are. The concept of Satan has often been used by men and women, and the Church, as an excuse for evil behavior, and a way of placing the blame for behavior on something outside of themselves. The issue of how much responsibility an individual has over his or her behavior has always been a complex one and much debated over the centuries. Unfortunately Freud didn't help matters with his concept of a powerful, uncontrollable Id, which had the effect of relieving man of some of the responsibility for his behavior. There is very little difference between blaming Satan and blaming the Id. I think that the Existentialist have it right when they say that we as adults, unless we are truly psychotic, must be held responsible for our behavior, and trying to hide behind a label, Satan, or the Id is no excuse.

Several years ago I went with my daughter to Krakow in Poland, and from there we took a train to Auschwitz where we toured the concentration camps. It was a surreal experience to see on display the piles of eye glasses, shoes, and mounds of human hair that the Nazi's had gathered from their victims. We took a bus to a huge field outside of town and toured the larger concentration camp at Birkenau. This is where the railroad cars came to a stop and the women, children, and old men were taken across the road to the crematory chambers and gassed with Kylon C, and their bodies burned to destroy the evidence. The rubble of the concrete roofs of the crematories and the chimneys are still there, like some kind of massive ugly sculpture to evil destroyed. What happened at Auschwitz and Birkenau is an example of great evil. The Nazi's had all the power and the innocent Jews had none. You would think that the world would have learned something from Auschwitz, but unfortunately it hasn't. Genocide is still rampart. Innocents are still being slaughtered. There is still much evil being committed in the world today.

We would like to think that we are far removed from our ancestors who roamed the savannahs of Africa, killing and pillaging a rival tribe at a coveted water hole; but in many ways we are not as far removed from those ancestors as we might think. We have but to pick up a newspaper and read how much evil man is committing each day to realize that man still has a long way to go. Not too long ago I was reading *The Dawn of Conscience,* by the Egyptologist, James Henry Breasted, written in 1933, and in the front of his book he quotes a passage from Ralph Waldo Emerson, who wrote in his *Essay on Politics* the following:

We think our civilization near its meridian, but we are yet only at the cock-crowing and the morning star. In our barbarous society the influence of character is in its infancy.

Little has changed since Emerson wrote that almost a century ago.

Depression, the Emotion of Despair and Hopelessness

Depression, the dark, helpless, hopeless, emotion of despair, that saps the energy and ambition of all who suffer from it, without question is the most devastating and deadly of all of the emotional illnesses, both in terms of morbidity and mortality. It takes an enormous toll in lives through suicide, but also through the daily emotional and mental drag that someone with depression has to suffer with on a daily basis. Research within the last few years has shown that it also takes a toll in other ways. We now know that individuals with depression are more likely to die of a heart attack, because studies have shown that depression makes the blood stickier, and more likely to clot in one of the coronary arteries, which can lead to a heart attack.

We often hear that depression is caused by a *chemical imbalance*, but lest we think that this is a new concept, Hippocrates also thought depression was caused by a chemical imbalance over two thousand years ago. For him it wasn't serotonin that was out of balance, it was an imbalance in one of the body's four humors. He thought that depression was caused by too much black bile, and it is from this that we get the term melancholia, because "melan" means black, and "cholia" means bile.

The etymology of the word depression comes from the Latin and means to push down. The first reference to depression as a state of mind was in 1425, but it was not formally used in psychiatry until 1905, during Freud's and Kraepelin's era. Some of the word and phrases we use to describe depression are words like discouraged, downcast, dejected, despondent, dispirited, down, glum, grim, down in the dumps, low spirited, morose, pessimistic, destitute, bleak, gloomy, melancholy, helpless, hopeless, despair, dark and black mood, woeful, dismayed, and loss of confidence.

Take note that I did not list sadness as one of the words to describe depression, because sadness and depression are two entirely different emotions, and it is important to make a distinction between them. Sadness is the emotion of loss, and it's normal to feel sad when we lose someone or something of value. Depression on the other hand is the emotion of helplessness and hopelessness, and is not the same thing as sadness. The confusion that exists about these two emotions is an old one and goes all the way back to the ancient Greeks. They felt that depression was sadness *without a cause,* and true sadness was sadness *with a cause,* and over the ages this explanation for depression has persisted. Today even the DSM accepts this explanation for depression, when it lists a sad mood as one of the symptoms of depression.

In my attempt to arrive at a rational explanation of the emotions, I start with the premises that all of the primary emotions, including depression, are normal when they occur at the right time and in the right amount. When I tried to fit depression into this paradigm I initially had some difficulty coming up with a set of circumstances in which depression was normal. I remembered a book which was written back in the 1970s that made the point that for our ancestors depression must have had some survival value. The title of the book was *When I say no, I feel guilty.* It was written by Dr. Manuel Smith, a psychologist, and though his book was mostly about assertiveness training, in the section on depression he wrote:

For our early ancestors, depression was a beneficial state when they had to put up with a period of harsh conditions in their environment. When things got rough, they had to withdraw to retrench. Our early ancestors, who got depressed and just sat around during very frustration times, were more likely to conserve their resources and energy. In doing so, they increased their chances for survival until better times came along.

Dr. Smith's comments on depression sounded very much like hibernation, so I began looking at the research on hibernation in animals to see if there might be clues to depression in humans. I ran across an article in the eighties by a researcher at Chapel Hill, N.C. who was doing work on hibernation in animals, and through one of his papers I came across the name of Dr. Peter Oeltgen at the University of Kentucky in Lexington who was doing research on a hibernating factor in ground squirrels. After reading several articles by Dr. Oeltgen, I called him and we discussed some of my ideas about depression and hibernation. Dr. Oeltgen's research showed that there is a specific factor in the brain of hibernating ground squirrels which causes them to hibernate. When this factor is injected into non-hibernating ground squirrels, they immediately begin to hibernate. It was also shown that the hibernating factor could be blocked by naltrexone, an opioid antagonist, which meant that the hibernation factor was a member of the opioid family of neurotransmitters, the same family that includes morphine and the endorphins.

I reasoned that if there is a specific factor that causes hibernation in animals, there could also be a specific factor in humans which causes depression; and if depression and hibernation are not the same thing, at the very least they must be similar. Hibernation enables animals to survive long periods in the winter without food or water, and during hibernation an animal's heart rate slows to four or five beats per minute and breathing slows to only a few times per minute in order to conserve resources. It seems reasonable to assume that at some point in our evolution some type of hibernation-depression center was built into our brains to help us survive helpless, hopeless circumstances. In the wintertime, when there were long periods of deprivation, our ancestors must have survived by crawling up in the back of a cave, pulling a bear skin rug over themselves, and sleeping for days until things got better. I read in a recent article in the New York Times that as late as the 19th century in northern Europe, [and that wasn't that many years ago in evolutionary terms], many peasants crawled up with their animal and slept through most of the winter, only waking once a week or so to stoke the fire.

I believe what we call depression comes from a center built into our brains to serve an evolutionary purpose, and is mediated by a specific depression neurotransmitter. Its original purpose was to help our ancestors survive helpless, hopeless circumstances, and it's still sitting there waiting to become active if we need it, and though it's rarely needed in today's world, under some circumstances it can still save us. Several years ago there was an earthquake in Turkey in which thousands of people were killed, and in the aftermath rescuers started searching through the rubble for victims who might still be alive. As the days went by the number of survivors became fewer and fewer. The last victim was found alive ten days after the quake in the rubble of a house that had collapsed on him. During the time he was trapped he just lay there quietly, without water or food, unable to move, but when he was finally rescued he was still alive. I believe this man survived because his depression mechanism kicked in, slowed everything down, and saved his life. He truly was helpless and hopeless. If he had struggled and exerted all of his energy, more than likely he would have died; but he lay quietly, conserving his energy, in a state of natural depression, waiting for help to come. When help finally did arrive, had enough left to come back. In the recent Haitian earthquake, incredibly one girl was found live 15 days after the quake, trapped in rubble. And the last victim, a man, was found alive twenty seven days after the quake.

Another example of normal depression occurred several years ago when a girl from Oregon left a party and said she was going home. When she didn't come home that night her family called the police, and they started looking for her. After several days her family and the police gave up searching, because they felt that she was either dead, kidnapped, or had run away. But the mother of one of her friends had the feeling that the girl was still alive, and kept looking for her. Each day she would go out and search along the roads that the girl might have been on. On the eighth day she thought she saw a car some two hundred feet below the highway in a ravine. She got help and went down into the ravine and saw that it was the girl's car. She expected to find the girl dead, but when she knocked on the window the girl answered her. She was still alive eight days after the accident. She had a head injury and had to be hospitalized, but she eventually recovered. I think this girl also survived because her depression mechanism slowed everything down, enabling her to survive. This is another example of *normal* depression. She was truly helpless and hopeless, but she survived because her depression center worked the way it's supposed to and saved her life.

Although a specific depression factor has not been found in humans, I suspect that the human depression factor, like the hibernation factor in ground squirrels, also comes from the opioid family of neurotransmitters. Evidence that there is a connection between depression and the opioids comes in part from the fact that until the tricyclics was introduced in the late 1950s, opium was the only drug that had any success in the treatment of depression since Paracelsus first used it in the 16th century. Dr. Emil Kraeplin, one of Freud's contemporaries, wrote a paper describing its use in depression in 1910. Opium was eventually abandoned when the tricyclic antidepressants were introduced, but the connection between the opioids and depression has not entirely disappeared. Within the last few years there have been a number of studies involving the use of opioids to treat patients with resistant depression. In one of those studies patients were treated with buprenorphine, an opioid agonist-antagonist, and some of the patients in the study showed remarkable improvement. Some said it was the first time that any medication had ever helped their depression.

In my own practice I have one patient with treatment resistant depression who had been tried on multiple conventional antidepressants with only minimal success. When I first saw her she was on an antidepressant, but still had significant depression. She also had chronic back pain which had not responded to other opioids, but when I added a small dose of methadone to her regime, she said for the first time since her depression began as a teenager, her depression was better and she felt more "normal." Though she still had some back pain, she felt the primary effect of the methadone was not on her pain, but on her depression. Her dose of methadone is very small and has remained stable. She comes in once a month for her prescriptions as instructed. She now has a job, whereas before she was unable to work. A search of the internet reveals that this is not a unique case and a number of studies have shown similar results with opioids in the treatment of resistant depression.

The use of the opioids to treat depression is not without controversy. Some would argue that the antidepressant effect of opioids is due to its euphoric effect, but when I talked with my patient about this there seems to be something else going on. The patient feels that the antidepressant effect of the methadone is not just due to the relief of her pain; because she still has some pain, but thinks instead that the methadone has a direct effect on her depression. I speculate that the antidepressant effect of methadone and the other opioids is due to an antagonistic effect on a depression neurotransmitter that is also an opioid.

An even closer connection between depression and hibernation is a condition called Seasonal Affective Disorder, or SAD, which incidentally I think is misnamed, because as I pointed out earlier sadness and depression are two different emotions. SAD is a condition which causes an individual to develop depression in the winter time, when the days are short, and the amount and intensity of light has diminished. I believe that SAD is caused by a depression-hibernation center which is overly sensitive to light, and becomes active in the wintertime. It starts pouring out depression-hibernation chemicals in the fall of the year in an effort to slow the individual down for the wintertime, just like it did thousands of years ago, when it enabled our ancestors to survive the winter. We know that SAD responds to conventional antidepressants which can turn off, or block, depression neurotransmitters, but it also responds to intensive light therapy which presumably also turns off the depression center. It would be interesting to see what effect naltrexone, the opioid antagonist that works on hibernating ground squirrels, might have on patients with SAD. I tried to get Dupont, which makes naltrexone, to become involved in such a study several years ago, but they told me they weren't interested.

The DSM list irritability as one of the symptoms of depression, but this only leads to more confusion about depression. Irritability is a form of anger, and though anger is often seen with depression, just as sadness is, it is a separate emotion. When someone has both depression and anger, it is not because the anger is a symptom of their depression, as the DSM states, it is because the individual has two different emotions at the same time. It is very important when this is the case that the therapist recognize and deal with both emotions, because if the anger is ignored there is the possibility that the individual will turn their anger on themselves, and sometimes on others as well, all too often with fatal results. Teenagers very often have a combination of anger and depression, but when they are put on an antidepressant like the SSRIs, the SSRI will sometimes make the irritability and anger worse before it helps the depression. Some of times these patients wind up killing themselves, and tragically they sometimes kill others as well, which is what I think happened at Columbine and Virginia Tech.

If my assumption is true and we all have a depression center, and depression is sometimes normal, what causes abnormal or "clinical" depression? I believe there are two causes for abnormal depression. One, an individual doesn't really have to be helpless or hopeless in order to have depression, they just need to see themselves that way, which is a cognitive problem. If a child is constantly put down, is told that they are no good and will never amount to anything, unless that programming is changed, when the child grows up they will more than likely continue to see themselves as helpless and hopeless. This will cause their upper brain, or superego as Dr. Freud called it, to send a steady stream of helpless, hopeless messages down to the depression center, causing the naïve depression center to put out depression neurotransmitters and cause the individual to feel depressed. This type of depression is solely a cognitive problem, and the type of depression that Drs. Aaron Beck and Albert Ellis began treating with cognitive therapy years ago. The goal of cognitive therapy is to reprogram the individual's upper brain and stop it from stimulating the depression center with helpless, hopeless messages. Sometimes medication is necessary to help get the depression center to calm down, even when the primary problem is one of cognition. If psychotherapy, or reprogramming, is successful medication can often be stopped.

The other cause for depression has nothing to do with programming or cognition; but instead is due to a malfunctioning depression center. Just as the anxiety center can malfunctions to cause panic attacks and PTSD, the depression-hibernation center can malfunction and put out depression neurotransmitters without cognitive input. Classic examples of depression-hibernation center malfunction are major depression and the depressive phase of bi-polar disease. When the cause of depression is depression center malfunction, the patient will feel depressed for no apparent reason. It won't make sense to them, and try as hard as they can to think their depression away, their efforts will be fruitless. They can't make their depression go away by thinking about it, any more than a diabetic can make their blood sugar fall by thinking about it. What's important for patients and therapists to understand is that when the patient's depression is the result of depression center malfunction, no amount of psychotherapy will help. It takes medication to treat this type of depression, and in rare cases ECT. This type of depression is also not usually the type of illness that can be cured, but like the insulin dependent diabetic, it can only be controlled by medication, and most of the time this means a lifetime of medication. Some patients resist taking medication on an indefinite basis, but if someone had diabetes and is told they have to take insulin the rest of their life, they wouldn't hesitate to do so. Patients with major depression and bipolar disease need to be educated to think the same way as someone with type one, insulin dependent diabetes, that they have a condition which cannot be cured only controlled.

Some psychologist and philosophers, including Jean-Paul Sartre and the late Dr. Robert Solomon, a philosophy professor, have said there has to be *intentionality* about the emotions, and that the emotions are always about something, that they serve some purpose; but this is not always true, especially where the primary survival emotions are concerned. The primary survival emotions come from centers in the brain that can malfunction, and act independently of the upper brain and cognition. When someone has bipolar disease and feels depressed, or manic, intentionality has nothing to do with it.

Whether the cause of the depression is poor programming and poor cognition, or a malfunctioning depression center, depression carries with it not only the morbidity of a life filled with misery, but a mortality risk that ranks with heart disease. Depression is an illness that can be treated just like any other illness, and it has absolutely nothing to do with a weak will. Fortunately we have numerous medications that can treat it, and treat it well. No one should ever be afraid, or ashamed, to admit that they have depression and deny themselves treatment. I have seen countless thousands of individuals whose lives have been transformed after their depression was diagnosed and treated. Sometimes it is necessary when treating depression to try a number of different medications before finding one that works and doesn't have significant side effects, but neither the patient nor the therapist should ever give in to the helplessness and hopelessness that is so much a part of this disease, and which sometimes makes it so difficult to treat. If the pharmaceutical industry can turn *impotency* into *erectile dysfunction* and make it acceptable, maybe they should start calling depression, *hibernation center malfunction syndrome*, and the treatment of depression could also become more acceptable.

Manic, the Emotion of Self-Confidence and Hope

In his Second Book of Rhetoric, Aristotle expressed the idea that many of the emotions have opposites, and the emotion which most philosophers, psychologist and psychiatrist would put opposite depression is happiness. But if depression is the emotion of helplessness and hopelessness, the emotion which is its opposite is the emotion of self confidence, self assurance, and hope. We don't have a very good term for this emotion and the term we use now is almost always thought of as being abnormal or crazy, but it is only abnormal when it occurs at the wrong time, and in the wrong amount. The emotion I am referring to is mania, and some of the words and phrases that we use to express different amounts and degrees of this emotion are words like exuberance, focused, full of energy, single minded, can't slow down, enthusiastic, confident, over the top, excited, frenzied, nutty, full of one's self, crazy, euphoric, omnipotent, hopeful, overconfident, and flight of ideas

The word mania comes from Latin and entered the English language around 1385, where it meant madness or insanity, and when we say someone is stark raving mad, it usually means that they are either in a highly manic state, or are schizophrenic. When an individual is in a full blown manic state they will lose all sense of responsibility, and do foolish or dangerous things which they may later regret, or not even remember. They can have a sense of omnipotence, often spend money unwisely, and run up huge bills on their credit cards, all with a total disregard for how they are going to pay them off. They may act out sexually, can have tremendous amounts of energy, and can go for days without sleep.

But if mania is one of the primary survival emotions and was built in, how much mania is normal? I believe that each of us has a manic center or manic area in our brains, and when it is working properly, it will put out the amount of manic chemical that matches the task that we need to perform. When we get up in the morning we need a little squirt of manic chemical to get us going. It's what gives us the confidence and energy to do what we need to do that day. If the depression chemical is the neurotransmitter of helplessness and hopelessness, and wants to keep us in bed all day, then the manic chemical is the neurotransmitter of hope and confidence, and is what gets us out of bed to begin the day's task. Caffeine is probably a mild manic stimulant, which explains why Starbucks is so popular. Stimulation of the manic center by drugs more powerful than caffeine is also a very popular, but illegal, pastime. Many of the popular street drugs including cocaine, ecstasy, methamphetamine, and other stimulant drugs, all have their effect on the manic center. When I have talked to patients who have abused cocaine, and who also have bipolar disease, they tell me that the "high" produced by cocaine is very similar, if not identical to the high that they have when they are in a manic state. The popular party drug ecstasy also has its effects on the manic center; and though it's effects are not as strong as cocaine's, they are more prolonged. The general assumption among users is that ecstasy is a relative innocuous drug, but there is strong evidence to the contrary. It too can be very addictive.

The brain and body don't like to be thrown off balance, so when someone stimulates their manic center through drugs, their depression center becomes active at the same time in an effort to restore homeostasis to the system, and balance out the excessive manic neurotransmitter. When the effect of the stimulant on the manic center wears off, the manic center shuts down, but the depression center doesn't shut down right away and this causes the individual to go through a period of rebound depression. If the manic center is stimulated chronically, such as with cocaine, it begins to hypertrophy, becomes larger and stronger and more unstable; and in an attempt to keep things balanced, the depression center hypertrophies as well. This results in two overgrown and unstable emotional centers that will cause an individual to start going through periods of mania and depression just like someone with bipolar disease. Tragically the depressive phase that can occur with cocaine rebound often leads to suicide, especially in young people.

Some Additional Notes on Bipolar Disease

The most common form of excessive manic neurotransmitters, not related to drugs, is due to the manic phase of bipolar disease, and when the manic phase of this illness is not so severe that it results in psychosis, it can lead to a period of great productivity. Not surprisingly many great writers have had bipolar disease. A classic example is Ernest Hemingway. He would have periods of high productivity when he was either in a manic, or hypomanic state, but his highs would eventually be followed by lows that would carry him into the depths of depression. It was while he was on a hunting trip in Montana, when he was in one of his depressed moods, that he took his hunting rifle and ended his life. Tragically, his granddaughter Margaux also took her life as well. .

Bipolar disease has one of the highest rates of suicide among all of the mental illnesses. The tragedy is that bipolar disease is a very treatable illness, yet countless thousands of people go through life trapped on an uncontrollable see-saw of emotions over which they have absolutely no control. Psychotherapy is worthless in bipolar disease; the only thing that helps is medication.

The DSM-IV divides bipolar disease into several types. The first is the most severe form and is called bipolar I. The second form is less severe and is called bipolar II. There is a third type called rapid cycling bipolar disease, which is characterize by very rapid mood swing, a condition in which the patient can go from feeling manic, to feeling depressed very quickly, sometimes in a matter of minutes. This is probably due to two very unstable manic and depression centers that oscillate very rapidly. There is another condition called the mixed state, where someone can feel both manic and depressed at the same time. To my knowledge there is no current theory that explains the mixed state, but it can be explained if we assume that when this occurs the two centers are producing their respective neurotransmitters at the same time, causing an individual to feel both emotions at the same time.

When treating bipolar disease it is often necessary for the physician to try a variety of medications until the right combination is found, and even then the medication may have to be adjusted over time. I have one patient with bipolar disease who I have been treating for over thirty years who has taught me something about the disease. She came to me originally in her twenties when she realized that something had to be wrong when in one of her manic moods she took off all her clothes, got into her MG with the top down, and drove out to a local lake to go skinny dipping. She also had periods of severe depression that would come on for no apparent reason. When I first treated her, I put her on a combination of Wellbutrin, an anti-depressant, and Depakote, a mood stabilizer. It didn't take long for her moods to stabilize, and with the exception of a few times when her moods were a little off, she had done well, and nothing like they were before she began treatment. She comes in once or twice a year for medication refills, has never been hospitalized, and lives a normal life. Over the last ten years or so, she has learned she can control her emotions by taking Wellbutrin alone. Most of the time on a medium dose of Wellbutrin her moods are normal and match the day's activities, but from time to time she will feel the depression starting to creep back in, and when she does she will increase the Wellbutrin for a few days, until her mood returns to normal, and then go back to her regular dose. Likewise when she feels the racing of mania starting to creep in, she will cut back, or stop, the Wellbutrin for a few days until the mania recedes, and then she will resume her normal dose. She has taught me that some patients with bipolar disease may need to adjust their medication in order to keep their emotions within a normal range. But in order for a patient to do this it is essential that they understand the nature of their disease, and this is really no different than teaching a patient with diabetes about their blood sugar, so that they can vary their insulin dosage to keep their blood sugar within a normal range. The most important thing that a physician can do for the patient with bipolar disease, apart from medication, is education, education, education. The patient needs to be taught that their disease is the result of two emotional centers that don't work right, and what medication it takes to keep their moods normal. To me this is not so much psychotherapy as it is patient education.

There are two caveats in treating patients with bipolar disease. Sometimes a patient will stop their medication because they feel they can control the way they feel without medication, but they soon find out they can't. Another problem is a patient will stop medication in order to bring on a manic high. In either case, stopping medication can be very costly indeed. They will either go into a high that can result in foolish and dangerous behavior, or into the deeps of depression that can result in suicide. With proper education, discipline, and medication, the vast majority of individuals with bipolar disease can live completely normal lives.

Sadness, the Emotion of Loss

Sadness is very often confused with depression. Physicians and philosophers have long considered depression as being a form of sadness *without cause*, and true sadness as being sadness *with cause*. The DSM even today lists a sad mood as one of the symptoms of depression; but depression and sadness are two entirely different emotions, and saying that sadness is one of the symptoms of depression is wrong. A rational theory of the emotions is based on which emotion would normally follow a particular circumstance or event; and the normal emotional response to the loss of something, or someone important to us, is sadness. If the importance of the loss is small, then the amount of sadness should be small, but if the importance of whom or what is lost is great; then the amount of sadness should be great. Aristotle recognized this connection many years ago and wrote that the amount of sadness was normal when the amount of sadness was proportional to the loss.

Patients rarely understand the difference between sadness and depression. When asked they usually say that sadness is something that goes away, but depression lingers on; and when asked to explain what kind of an event causes someone to feel sad, most patients are at a lost to explain what actually causes sadness. They will often say it is when something bad happens to you, but they are not usually able to say what kind of a bad thing that is, they just say something bad. When I point out to them that sadness is always the result of a loss of something or someone important, they are finally able to make the connection, but it is rare in my experience to find a patient who is able to make the connection between sadness and loss the first time.

Sadness, and its opposite, happiness, the two non-survival emotions, probably do not have specific centers in the brain, but instead depend on a blend of neurotransmitters that are the brain's response to gains or losses in the external world; and because of this they probably are not subject to emotional center malfunction the way the primary survival centers are.

The meaning of the word sad has changed greatly over the years. It's meaning before about 1200 meant to be satisfied, or to be heavy with something. Gradually this evolved into being weary of, or to be tired of something, and eventually into sorrowful or unhappy. Webster's New World Dictionary defines sad as having or expressing low spirits, unhappy, sorrowful, causing dejection, very bad, deplorable----but interestingly, it does not mention sadness as being caused by the loss of something or someone, so even Webster fails to make the connection. Some of the words that we use to express sadness are sorrow, unhappiness, blue, mournful, desolate, devastated, brokenhearted, grieving, heartsick, woe, disconsolate, doleful, forlorn, heavyhearted, low, pensive, tragic, bereavement, melancholy, dispirited, hurt, and crushed. Note that some of these words are the same words that we use to describe depression, so it is not surprising that the two emotions are often confused in people's minds. Sadness and grieving is without question one of the most painful emotions that we have to face in life. But to grieve and feel sad when we lose something, or someone important to us, is normal.

My wife, Merrie, lost her mother to colon cancer when her mother was only in her late fifties. She and her mother were very close, and the loss of her mother was almost more she could bear. She was devastated and found her mother's death very difficult to understand or accept. In an effort to understand her loss, and be able to deal with it, she did something that I thought was very brave. Merrie joined a hospice class and became a hospice volunteer. She decided to meet grief head on and not run from it. She took classes in death and dying, and when the classes were over she was assigned an elderly lady near our home who was dying of cancer. Merrie would go by and see her every few days. They would talk about their children, their love of flowers and other things in their lives, and as time went by her patient became weaker and weaker. One day Merrie got a call that she didn't need to come any more. I was very proud of her for coming face to face with loss, something that most of us would rather not face at all.

The DSM assigns a figure of two months for the time of grieving after the loss of a spouse; and goes on to say that if the grieving last longer than two months, depression is involved. This is another DSM time frame that is totally unrealistic, and assigning a two months period of grieving after the loss of a spouse is purely arbitrary and unreasonable. The normal process of grieving can vary tremendously, and the only way to determine if the patient is grieving normally, or has slipped into depression, is through a careful dialogue with the patient, taking into consideration the culture and the nature of the relationship with the one who is lost. Having said that, it is not uncommon for a surviving partner to have both sadness and depression. This is very often the case when the remaining spouse had been very dependent on the spouse who has just died. The remaining spouse often feels that without his or her mate they are totally hopeless and helpless, which leads to depression. In grief therapy it is important for the therapist to help the patient separate the two emotions, because no amount of anti-depressants is going to make the grief go away, but it can help the depression. The only treatment for grief is time.

I had a patient several years ago whose life events illustrated this in a very poignant way. I had been taking care of her and her husband for a number of years. They adored each other. He was very successful and traveled internationally. Sometimes she would go with him, and if he was away and she was at home, he would call her frequently. They had two boys, one was in high school and the other was in college. She and her husband both went to the gym and worked out regularly, neither smoked, were conscious of their diets, and came in for regular checkups. They both appeared to be in excellent health. One day while they were working out in the gym the husband collapsed. An ambulance was called, and he was rushed to the hospital, but he was dead before he got to the ER. The next day she came to the office in tears. She was devastated at the loss of her husband, and I sat with her for a long time, and we talked about her feelings of helplessness and hopelessness, and her feelings of depression. She felt that she could not go on without him. We also talked about the overwhelming sadness that she felt and in one of those first visits I introduced her to the concept of the difference between the helplessness and hopelessness of depression, and the sadness of losing someone you love. I explained to her there was a natural process of grieving that she would have to go through and nothing could take that away from her. I also told her that I understood her feelings of helplessness and hopelessness leading to her feelings of depression over the loss of her husband. I put her on an antidepressant and over the next several months she came in on a weekly basis. Slowly her depression lifted, and her grief and mourning began to resolve. She found a job and went to work, which also helped. As difficult as it was, she and her two boys went on with their lives.

She eventually came off of the antidepressant, and though she still missed her husband, she had moved on. About a year later I saw her name on the appointment book and I assumed that she was coming in for a regular visit. When she came into my office, she burst into tears and I said, "What's wrong?" As she fought back the tears, she told me that on the previous weekend her oldest son had gone up to visit his girl friend at a college near Asheville. He was riding along on I-40 at night and apparently failed to see that a portion of the highway had been closed for construction, and ran head-on into a piece of heavy equipment, and was killed instantly. The tears welled up in my own eyes and I slowly began to sob with her.

It's not often that someone has to bear the burden of two significant losses within such a short period of time. We both knew what she was going to have to go through. It would not be the heaviness of depression this time; it would be the almost unbearable pain and grief at the loss of a son. I couldn't take that away from her. She and I both knew that no amount of antidepressants would help. We both knew that only time would blunt the pain of her loss. She came back to the office often and we talked about her losses, and why tragedy sometimes seems to strike some so unfairly. I didn't have an answer. The only thing that I could say is that sometimes the weather of our lives is stormy, and seems so unfair. If you were a tree, I said, it would be like you had been hit by a hurricane in September, and had some of your limbs torn off, only to be hit again by an ice storm in January, and had more of your limbs torn off by the heavy ice. I said there is no explanation for why some of us have to endure times which seem so unfair. I told her that I didn't think that it was fair to blame God, as some so often do, because I think God is just as sad as we are.

Over the next few months her grief began to subsided and she was able to return to work. Sometime later she met a man from New England. They fell in love and got married. With some reluctance she moved to a small town outside of Boston, and when she came back to Raleigh, she would either call, or come by the office, and let me know how she was doing. Apart from a few minor problems adapting to a different environment, she was getting along well. Sometime later I was walking along the beach near Wilmington, N.C., when I ran into her with a group of women friends. She had come down from New England for a vacation and seemed happy and content. That was the last time that I saw her. I did hear later that the second husband had also died.

Another emotion that often complicates loss and sadness is anger. This is especially true when the loss of a loved one actually is unfair, or is perceived to be unfair. When someone has sadness and anger, resolution of the normal grieving process can be very difficult, and the only thing that can help resolve the grief is forgiveness, which is often the most difficult thing that someone ever has to do. Someone who has both grief and anger needs to understand that by continuing to harbor the anger, the only one the anger really hurts is themselves, not the one the anger is directed toward. It's like taking poison and expecting the other person to die.

I once had a family whose daughter was addicted to drugs and alcohol. The mother and father did everything within their power to try and help her. She had been admitted to treatment centers numerous times, but despite all of their efforts she continued on a path that eventually led to a tragic end. One night she went to a drug house, presumably to buy drugs, and while she was there, someone came in to rob the dealer and during the course of the robbery she was shot and killed. The parents were understandably devastated. Not only were they grief stricken at the loss of their daughter, they were filled with rage at the man who shot her. He was captured, eventually put on trial, convicted, and sent to prison.

Following this tragic event I initially saw the parents in the office on a regular basis in counseling. At each session the father's grief and rage continued, and I knew that it was hurting him. He had high blood pressure and it kept going up, and I knew that as long as his rage continued, he likely wouldn't get any better. It could only hurt him, maybe even cause his death. One day as we were talking, I said, Frank (not his real name), if you are ever going to get better you are going to have to do one of the hardest things that you could possibly have to do. I said you are going to have to forgive the man who shot your daughter. At first he said he didn't think he could do it, but over the next few weeks and months he began to get better. I'm not sure that he ever truly forgave the man, but I think he understood what I was talking about. Forgiveness under such circumstances is one of the most difficult things that we humans could ever have to do.

I remember when the late Pope John was shot and almost killed a number of years ago, one of the first things that he did after he got out of the hospital, was to ask permission to go to the prison cell where his would-be assassin was being held. He did this with the express purpose of telling the man that he forgave him. He knew that without that act of forgiveness, he would have to carry in his own heart a burden he knew would only hurt him and not his assailant. Forgiveness is rarely easy, but sometimes it is necessary to save our own souls.

Sadness is one of the most painful of all human emotions, but it is not the same as depression. Sadness is different, and it's important to know the difference.

Happiness, the Emotion of Gain

If you ask most people what they want out of life, they'll say they just want to be happy. And if you ask them what it would take to make them happy, they will name things like money or health, and if you ask them what actually causes happiness, most of the time they are at a loss to explain where happiness actually comes from. Happiness is the emotion that is the opposite of sadness, and if sadness is the emotion we feel when we lose something of value, happiness is the emotion we feel when we gain something of value. The gains of happiness can either be tangible, such as winning the lottery, the birth of a child, or an A on a test; or the gains of happiness can be intangible, such as when "my" team wins the championship, or when "my" child learns how to ride a bicycle, or when I see a beautiful sunset.

Happiness is not the same thing as mania, with which it is often confused, just as sadness is often confused with depression. Happiness comes from the neurotransmitters of gain, and mania comes from the neurotransmitters of self-confidence and hope. Euphoria is sometimes used to describe both of these emotions, but euphoria is also used to describe the feeling that comes from excessive endorphins or opioids, which illustrates one of the difficulties we have putting the emotions into words.

The word happy entered the English language sometime around the year 1380 and was derived from the English word *hap*, which meant fortune or chance, to which a "py" was added. Some of the words that we use to express happiness are words like good cheer, bliss, enjoyment, good humor, gladness, good spirits, jovial, laughter, pleased, delight, ecstasy, elation, joy, jubilation, mirth, pleased, exhilaration, euphoric and exuberance. All of these words describe various degrees of happiness, but it is important to remember that it is rare when our emotions are pure. Much of the time our emotions occur as blends, and occasionally as opposites. When sadness and happiness occur at the same time, we have a name for it, it is called bittersweet.

We also need to remember that there can be a dark side to happiness. Happiness like mania can be addictive. A very common form of happiness addiction is gambling, an addiction to the big win or gain. Though some have questioned whether it is possible to become addicted to gambling, studies have shown that it has all of the earmarks of the other addictions, including symptoms of denial and withdrawal. It also responds to the twelve step programs. Gamblers Anonymous is a program for gambling addiction and is very similar to Alcoholics Anonymous. Most large cities have one or more chapters, and in Las Vegas, not surprisingly, there are more than a dozen chapters.

Gambling addiction is probably more common than is generally recognized. My wife has a good friend whose ex-husband was addicted to gambling, and he would have lost everything, but for a mother who kept bailing him out. There are countless thousands of individuals who have gambling addiction, and if they don't recognize it as a disease and receive help, many of them will go on to lose everything they have, and some will eventually commit suicide. Gambling addiction can strike individuals from all walks of life, just as alcoholism can. Not too long ago one of the most respected voices of conservatism, William Bennett, came forward and admitted that he was addicted to gambling. This was after it was disclosed that he had lost huge sums of money at the casino tables.

People who say, all they want out of life is just to be happy, fail to recognize that in order to feel happy all of the time, it means going through life with a never ending series of gains, which is not really possible in this life. It is possible for most of us to experience much happiness during our lifetime, but if one wants to experience the most happiness in this life, the types of gains that will most often accomplish this are the intangible ones, like the smile of a ten months old granddaughter, the beauty of a sunset over water, a cool breeze on your face in the heat of summertime, an adagio from one of Brahms's symphonies, the delight of a soft bed to crawl into at night when you are tired, and the joy one feels when one of your children succeeds at a difficult task. These are the type of gains that are both priceless and mostly free.

Some Notes on Causes of Malfunction in the Four Primary Survival Emotional Centers

There are two main causes for primary survival emotional center malfunction. First, there is a strong genetic influence on the size, strength, stability, and sensitivity of the primary survival emotional centers. We know that bipolar disease runs in families and is a perfect example of two emotional centers run amuck. When I see a patient with an emotional problem, it is usually the case that someone else in the family has a similar problem. This is especially true if the patient has either depression or anxiety. In fact it is rare when I see a patient with either one of these problems that I don't get a positive family history. It is usually a parent or grandparent, sometimes an aunt or uncle. Often the patient will give a clear cut history of a family member that has a similar problem, but sometimes it is more subtle. The patient may tell me that the mother doesn't have an anxiety problem, but she takes Valium all of the time, or that the father drinks too much. I have found that alcoholism is a very strong marker for both depression and anxiety, especially in men. Men who have anxiety problems are especially prone to develop alcoholism because they start out using alcohol as a way to treat their anxiety, and go on to develop alcoholism. Men in general are also notorious for hiding their emotional problems, but in the last few years this seems to be getting a little better, and this is probably the result of better education about emotional illnesses, and the increasing acceptance of it by society.

When there is a strong family history of a particular emotional problem, such as anger or depression, education can be an important tool in helping the patient deal with an inherited problem. Once a patient understands that they have the potential for a particular emotional problem, they are usually more receptive in getting help, especially when they realize that it is not normal for them to feel depressed, angry or scared all of the time. Not too long ago I had one of my patients, who is a high school student, come to me and say he suspected that he was developing some of the same symptoms as his mother, who has bipolar disease. He said at times his mind would start racing, and at other times for no reason he would feel down in the dumps. He was only 15 years old at the time, but after evaluating him I agreed with him that he had made the correct diagnosis. I put him on one of the mood stabilizers, and a follow up showed that his racing thoughts had stopped and he no longer had periods of depression. I suspected at the time that he probably would have to stay on medication indefinitely, and I explained to him that this was no different than telling a young person they had diabetes and would have to stay on insulin the rest of their lives. In his case it would be his emotions, and not his blood sugar that I was trying to keep in a normal range.

Sometime later he came to the office for another problem, and I asked him how he was doing emotionally. He said he had come off of his medication after he learned that he could control his moods by running and exercise. I encouraged him at the time to be sure to come back to see me if exercise quit working. He is a bright young man and I think he understands the nature of his disease. I only hope if exercise quits working he will come back to the office for help and not let his disease run on unchecked.

Another cause for survival emotional center malfunction is the result of hypertrophy which can occur in childhood. Research has shown that the child's brain is very plastic and adaptable, and early childhood experiences can affect the growth and size of the primary survival emotional centers. If a child has a very stressful childhood, one or more of their survival emotional centers may become hypertrophied, the same way the muscles do with exercise. If a child grows up in hell, and is constantly beaten or abused, it is not surprising that the child will develop an anxiety center and/or a depression center that is large and overactive. Many children growing up in Iraq today will undoubtedly develop anxiety centers that are hypertrophied and overly sensitive, and have to deal with chronic anxiety, or panic attacks as adults. If a child is constantly told that they are no good, will never amount to anything, and are constantly put down, it is not surprising the child will not only be programmed to see themselves as helpless and hopeless, but will also develop a depression center that is large and overactive. If a child is raised in an environment where there is constant fighting and anger, the child's anger center may hypertrophy, and leave them with an anger center that is overgrown and sensitive, and one they will have to deal with as an adult.

Some Additional Notes on Why Patients Are Often Confused About Which Emotion They Are Actually Feeling

Patients often confuse one emotion for another, because our emotions don't carry any labels with them. When I am dealing with a medical illness I can draw a blood sample or get an x-ray to help me make a diagnosis, but I can't do that to help me identify an emotions. One day in the future I think we will be able to get help from the lab by having a patient's blood analyzed for the type and amount of the emotion that someone has, but for now we have to rely on observing the patient's posture, their appearance, the words they use to describe an emotion, their actions, and data from family members to get a rough idea of which emotion, or emotions, the patient has. Patients often mistake anxiety for depression, and vice versa. It is also very common for a patient to have several emotions at the same time.

I have found that it is helpful to educate the patient about the primary emotions when I first see them, even before we get into which emotion they actually have. I explain to them there are six primary emotions and each of them come from specific centers in the brain, and I hold up my hand and point to the little finger and say that one is anxiety. It is the emotion of that helps us deal with harm and danger. The next one is anger, aggression, and assertiveness, and I point to the ring finger. That is the one that helps us get what we need in order to survive and protect what is important to us. The middle finger is depression, the emotion which protects us when we are truly helpless and hopeless. The index finger is manic, the emotion of confidence, we need a little bit of it in order to get things done, but not too much, because too much can make us crazy. The thumb is sadness, the emotion we feel when we lose something, or someone, important to us. The last emotion is happiness, the emotion of gain, and I have to bring over the index finger of the other hand. These are the six primary emotions. Four were built in to help us survive in the outside world, but sometimes they don't work right. Sometimes they overproduce and cause a problem like panic attacks. Sometimes they just sit there and pour out their emotional neurotransmitter for no reason at all, which is what happens in bipolar disease when the manic and depression center are producing their neurotransmitters. The two non-survival emotions, happiness and sadness, are often confused with manic and depression, but they are different and not the same. I want the patient to understand the differences between the primary emotions, what each one means, where they come from, how they differ, and more importantly when an emotion is normal, and when it is abnormal.

I have also found it useful when I am talking with a patient about their emotions to avoid the use of the verb "to be." I don't say to a patient, you *are* depressed, or you *are* anxious. I use the possessive and say to the patient, you *have* depression, or you *have* anxiety; because I want the patient to understand that their problem is not something that they *are,* but is something that they *have,* and when they are feeling anxious or depressed it is because they *have* the neurotransmitters of anxiety or depression in their brains and bodies. This is one of the things that I learned from the French language. The French don't use the verb "to be" the same way we do, particularly when they are describing the emotions. They don't say, for example, I am afraid, they say *J'ai peu,* I have fear. When a patient has a panic attack it is because they *have* too many anxiety neurotransmitters in their system for the circumstance. It's a subtle difference, but I think a very important one. When I am talking to a patient who is addicted to alcohol, I don't say, "You are an alcoholic," I say, "You have alcoholism," a disease which can be treated. I think by making a distinction between *having* and *being,* the Existentialist make an important contribution to a better understanding of the emotions. Dr. Albert Ellis recognized this difference and actually wrote several books in E-Prime, a technique he used to avoid the use of the verb *to be,* but he eventually abandoned this form. I once wrote him a letter and asked him why he had abandoned this concept and he wrote me back and said he did so because people didn't understand it. But I think that if we keep this concept in mind, we can avoid some of confusion that exists about the emotions.

Sometimes when a patient comes to the office it turns out that the emotion they have is normal. When this is the case I reassure them that though they might not like the way they feel, under the circumstances the way they feel is normal; and the best thing they can do is to accept it, or change the circumstances if they can.

Some Additional Thoughts on Freud's Psychology and Addiction Disease

Freud was not the first to propose that the human psyche is divided into three parts. Many centuries before Plato had written essentially the same thing. He said that the human psyche, or mind, was a tripartite entity, and he likened it to a chariot driver struggling to control two horses pulling in opposite directions. One of the horses was dark, headstrong, and lustful; the other horse light colored and more reasonable. The light colored horse was trying to pull the chariot one way and the dark unruly horse the other, with the driver fighting the reins to keep the chariot going in the direction that he chose.

Freud also divided the human psyche into three parts. He said that one of those parts was hidden in the unconscious and was very powerful, not unlike the dark, lustful, and headstrong horse in Plato's psychology. Freud called that part the id, a word he borrowed from Latin where it means *it*. He said our drives and cravings came from the id, especially the sex drive, and he said that the id operated on what makes us feel good, a force he called the pleasure principle. The id in

Freud's psychology was a very powerful force and one to be reckoned with, if not feared. The second part of the psyche in Freud's psychology was not the white horse of reason in Plato psychology, but the voice of our parents in our heads. He called this part of the psyche the superego. He felt that the superego was restrictive, forever judging us, telling us what to do or not to do, and forever making us feel guilty. In between the id and the superego was the ego, which is the Latin word for *I*. The ego's task was to try and reconcile the restrictive, condemning superego on the one hand, and the powerful, lustful id on the other, both demanding to be satisfied. Behavior would be dictated by which one had the most influence over the ego, and how strong the ego was. If the superego was the stronger, it would dictate behavior that was mostly restrictive, don't do this, don't do that; but if the id was stronger, it would dictate behavior that was more permissive; go ahead do this, do that. This struggle has often been depicted in cartoons as an angel with a halo sitting on one shoulder whispering in one ear, and a devil with horns sitting on the other shoulder whispering in the other ear.

Modern research would show that the human psyche is far more complicated than either Plato or Freud realized, but they are probably right in saying that the mind is divided into three parts. What Freud called the superego, in today's parlance we would call programming, a concept that Freud was not aware of. (Freud was not that far off though. He reasoned that parental influence had to have been stored somewhere). Today we know that the programming we receive as children, as well as the programming we receive as adults, isn't necessarily restrictive or condemning and can be either positive or negative. It can come not only from parents, but from a wide variety of sources, including friends, experience, teachers, circumstances, books, TV, etc.

When Freud was developing his concept of the id he was also not aware of the many neurotransmitters that arise out of the lower brain, and in addition to the drives that Freud was talking about, like the sex drive, the lower brain also contains the neurotransmitters of the primary survival emotions, including the neurotransmitters of anger, anxiety, depression, and mania which can operate outside of cognitive control.

Part of the problem with Freud's psychology and psychoanalytic theory may not be so much his fault, as it was the fault of those who followed after him. Many of his followers took everything that Freud said as being the final answer, and they stopped looking for new solutions. Freud, on the other hand, was ever inquisitive and never stopped looking for new knowledge. If Freud knew then, what we know now, I feel certain he would have adapted this new knowledge to his theories. Freud's friend and biographer, Ernest Jones, wrote that Freud always felt that the key to emotional illness one day would be discovered in the laboratory, and I am convinced that one day Freud will be proven right.

One of the problems with Freud's psychology is that not everyone fits into his neat psychoanalytic mold. Not everyone has an overbearing and punitive superego. Not everyone has an id that urges them to do dangerous things, or a weak ego that can't say no to the id or stand up to an overbearing superego. Unfortunately many of his followers applied Freud's model of the psyche to everyone which often was a mistake. Certainly some individuals have a voice in their heads that constantly puts them down and condemns them, but this is not true for everyone; and not everyone has an id that urges them to do crazy things. When a therapist evaluates a patient, it is essential that each patient be evaluated on an individual basis, and the therapist can make a mistake in assuming that every patient has a superego bent on making them feels guilty, or an id which is powerful, dark, and hell-bent on driving them to destruction. Just as the strength of the superego or programming varies from individual to individual, the strength and cravings of the id and the output of one of the primary survival emotional centers also varies from individual to individual. It is only after an objective evaluation of the patient can a therapist say whether the patient's id is too strong, the superego too condemning, or whether it is one of patient's primary survival emotional centers that is malfunctioning.

A similar evaluation needs to be done before deciding about the relative strength of the ego. To assume that the ego is always weak, relative to either the id or the superego, is to make an assumption that may not be true. All too often in the past it was assumed by the analyst that the ego was always weak, when in fact the problem was not the ego at all, but the problem was a malfunctioning primary survival emotional center, and had nothing to do with a weakness of the ego. A weak ego is not what causes someone's emotions to go up and down like a yo-yo when they have bipolar disease, or sink into the depths of despair when they have major depression, or have the terrors of the night when they have PTSD.

Addiction Disease

There is one ailment where Freud was correct about the strength and power of the dark side of the id and that is in addiction disease. In addiction disease, the driving, craving, lustful forces of the lower brain, or id, take over control of behavior from a weak ego. The id starts calling the shots, and sometimes that literally means calling for shots. And there is another set of neurotransmitters that Freud didn't know about that should be included in the id. Recent research has shown there is a set of neurotransmitters in our lower brain called the craving neurotransmitters, and these play a critical role in addiction disease. The number and strength of the craving neurotransmitters will determine the intensity of a particular craving or need, and as the number of craving neurotransmitters increases, the urgency for the individual to do, or to seek a particular thing, becomes more intense. When there are only a few of the craving neurotransmitters present, the individual feels that I would *like* to have a particular thing, or would *like* to feel a certain way. As the number of the craving neurotransmitter increases, the individual begins to feel like I *need* a certain thing, or *need* to feel a certain way; and as the number of the craving neurotransmitter continues to climb, the individual begins to feels like I *have* to have a certain thing, or *have* to feel a certain way. At this point the individual begins to feel desperate for whatever it is that they want to feel, or think that they have to have.

When we are very young the craving-seeking neurotransmitters are very strong relative to the ego, and to observe the craving neurotransmitters in action one has but to go to the nearest toy store and observe a typical five year old. As children we have a tendency to want what we want when we want it. As we mature our upper brain, or ego, normally learns how to evaluate circumstances and decides when it is appropriate to let the craving neurotransmitters have their way, and when to say no to them. In addiction disease the ego, or upper brain, loses its ability to say no to the craving neurotransmitter of the lower brain, and the lower brain takes control of behavior. If someone is stranded in the desert without water, a high level of craving neurotransmitters for water would be normal, but if someone has the same level of craving neurotransmitters for another shot of heroin in order to get high, addiction disease is involved.

Contrary to what most people think about addiction disease sometimes the addiction is not actually to the substance that someone is taking, or to the act that someone is committing; but in reality the addiction is to one of body's own neurotransmitters that is being stimulated by the substance they are taking or the act they are committing. We know that two of the neurotransmitters involved in pleasure are dopamine and the endorphins, but there are number of others as well, including the manic neurotransmitter. If there is a neurotransmitter that makes someone feel good, or is pleasurable in some way, even if it is in a way that most of us would find bizarre, it is possible for someone to become addicted to it.

One of the issues that has never been settled about addiction disease is the issue of responsibility. If the id is as strong as Freud proposed, is the person with addiction disease responsible for his or her behavior? Jean-Paul Sartre felt that our conscious self, or ego, must ultimately be given the responsibility for behavior, whether the issue is addiction or actions during wartime. He disagreed vehemently with Freud about the strength of the id, and said that to blame bad behavior on the id is basically a cop out. There are times when it is permissible for the upper brain to give the lower brain what it wants, but this is only after the circumstances have been weighed, and the risk benefit ratio measured. If an individual is in pain from the throes of a kidney stone, then the upper brain would rightly decide it is permissible to get a shot of morphine. The addict on the other hand would decide to give in to the same amount of cravings from the lower brain, not to relieve pain, but to have the feelings of euphoria one more time.

Endorphin Addictions

The body has a store of natural painkillers called endorphins, and when someone is in pain, endorphins are released from various storage sites and fit into specific receptor sites scattered throughout the brain and body. When endorphins fit into these receptor sites, pain is relieved or partially relieved. It is not uncommon for a patient to come to the emergency room with horrific injuries, and when asked whether they are hurting or not, they say they have very little pain. This is because their own endorphins have been released to help relieve their pain, and when someone is in pain it means that their endogenous endorphin levels are inadequate to relieve the pain.

Doctors use opioids like morphine, codeine, Demerol, Oxycontin, etc. to relieve pain. These drugs are called endorphin agonist, because they will fit into the same receptor site as the body's own endorphins, and have the same effect in relieving the patient's pain. But if the level of endorphins, either endogenous or exogenous, rises above the level of pain, the individual will begin to feel euphoric, better that normal. It is to this feeling of euphoria that the person with narcotic addiction becomes addicted to, and has the craving to go back to. It has sometimes been described as "feeling no pain," and as one addict described it, it is like being "enveloped in the arms of God."

When someone is not in pain, it is possible to reach the euphoric state with a much lower level of an endorphin agonist than it is when someone is in pain. But whether it is when someone is in pain and takes enough endorphin agonist to rise above the pain to reach a state of euphoria, or when someone takes an endorphin agonist when they are not in pain and reaches a state of euphoria, once someone experiences the feeling of euphoria there is the risk of addiction and the desire to go back to that feeling. Individuals with narcotic addiction usually become addicted to opiates through recreational use when they are not in pain, but it is also possible for someone to become addicted when they are in pain. When a doctor prescribes an external endorphin, such as morphine or Oxycontin, it is important to remember that as long as there is not an attempt to completely relieve the patient's pain; the euphoric state can be avoided.

In either case, whether the euphoria is due to excessive opioids in the relief of pain, or the use of opioids through recreational use when there is no pain, the wisdom of the body soon recognizes a state of excessive endorphins and will begin to shut down the body's own endorphin factories. Eventually with continued opioid use, the body's endorphin factories shut down completely, and now the individual is totally dependent on an external source of endorphins to supply the body's basal level of endorphins, which is necessary for the individual to feel "normal". It is a common misconception that the heroin addict is always "shooting up" in order to get high. After using external endorphin for a period of time, the addict's own endorphin factories will shut down, and now they are totally dependent on a daily supply of external endorphins in order to maintain a basal endorphin level. Of the patients that I have treated in the office with heroin addiction, several have told me that it takes an average of three bags of heroin a day to keep withdrawal at bay, and this is apparently how much heroin it takes to supply a basal level of endorphins in order for them to feel "normal".

Once an addict's endorphin factories have shut down it now becomes very dangerous for them to try and reach a state of euphoria, because the amount of an external endorphin, such as heroin, that will produce the feeling of euphoria is very close to the level that will shut down the respiratory center, and many addicts die from an overdose in an attempt to reach the euphoria state after their own endorphin factories have shut down. If an individual with endorphin addiction stops using external endorphins, he or she will start going through withdrawal, a state in which the system is devoid of any endorphins. Withdrawal will last until the production of the addict's own endogenous endorphin factories returns to normal, and the length of time that it takes for endorphin levels to return to normal will be dependent on the length of time and the amount of external endorphins that they have been using. Many addicts quit on their own and go through withdrawal successfully, as was shown after the Vietnam War when many soldiers became addicted to heroin but kicked the habit when they came home. But withdrawal can be very uncomfortable and associated with a number of side effects including nausea, seizures, muscle cramps, etc. There are several drugs which can help the withdrawal symptoms including clonodine, a blood pressure medicine, and the benzodiazepines.

After the addict goes through withdrawal and their endogenous endorphin levels return to normal, the real danger now comes from recidivism, because once their endorphin production returns to normal it is easy to reach a state of euphoria again without the risk of overdosing. Unfortunately the craving to return to the feeling of euphoria drives many users to begin using again, and without the ego-strength of an upper brain that can say no to the craving chemicals of the lower brain, they remount a horse that will take them on a merry-go-round that goes from euphoria to hell, and back again. Often this ride ends in death.

One way to treat heroin addiction that avoids withdrawal is to use a substitute endorphin like methadone, which is a long acting endorphin agonist which can be taken by mouth. It will replace the deficiency of endogenous endorphins and prevent withdrawal, but as long as the addict takes external endorphins, their own endorphins levels will never return to normal. I have had users who were on methadone who told me that the way they feel when they are on methadone is just not quite the same as when they are clean and producing their own endorphins. Apparently anything other than natural endorphins feels somewhat artificial and is probably due to the difference between an artificial exogenous endorphin agonist like methadone, and the body's own natural endorphins. But taking a daily dose of methadone for some is better than obtaining a maintenance dose of endorphins on the streets, or going through withdrawal. Another effect of methadone is when someone is taking forty milligrams a day or better they can't get high off of heroin, because methadone will jam all of the endorphin receptor sites and heroin will have no place to fit.

Another treatment for narcotic addiction is to use a drug like naltrexone, an oral drug that is an endorphin antagonist. Its mode of action is to jam the endorphin receptor sites so that heroin, or any of the other endorphin agonists, can't fit into the endorphin receptor sites and prevents the addict from getting high. Another treatment for endorphin addiction is the use of a mixed endorphin agonist-antagonist called buprenorphine. It will also fit into the endorphin receptor site, and although it has a mild endorphin effect as an agonist, it also acts as an antagonist and will prevent another endorphin agonist from fitting into the site. One of the latest treatments for endorphin addiction is a combination of buprenorphine and naloxone. Naloxone is another endorphin antagonist, so this combination has several effects; a mild agonistic effect, combined with a double antagonistic effect.

Interestingly, buprenorphine has also been used to treat resistant depression, which adds to my suspicion that there is a connection between the depression factor and the opioids. As I pointed out in another section opium was used for many years to treat depression up until the tricyclics were introduced in the 1950s. I have had several patients over the years with chronic pain who reported that narcotics not only relieved their pain, but also their depression. I am convinced in some individuals the opioids have an antidepressant effect. The psychiatric literature has a number of studies that confirm an antidepressant effect of the opioids in some individuals.

One aspect of illegal drug use, often overlooked, is the frequency with which users are attempting to self-medicate. Many individuals start using narcotics in an effort to treat themselves for a variety of physical and psychiatric problems. The problem of self-medication by addicts was pointed out in a paper written in 1953 by Dr. A Wilker who said there were a number of reasons, besides getting high, that lead some individuals to start using illegal drugs. He found that some were trying to treat their anxiety, others were trying to treat their depression, and still others were seeking relief of pain. It is important for the physician who is treating narcotic addiction to look for an underlying cause of the patient's drug seeking behavior. It may well be that the narcotic user is not just trying to get high, but may be trying to treat an underlying medical or psychiatric problem.

Nicotine Addiction

The number one addicting drug worldwide is nicotine, and the one drug that I personally was once addicted to. When I was in medical school in Winston-Salem, home of the Reynolds tobacco company, salesmen from Reynolds used to come up on the wards and pass out samples of Winstons. No one told us at the time that the drug was addicting and I smoked off and on for the next twenty years. Periodically I would quit and every time I would start back, I would promise myself that I was only going to smoke three cigarettes per day. That would last about a week, and then I was right back to where I was before, smoking from a pack to a pack and a half a day. It was not until I remarried in the late 70s that my new wife insisted that I quit smoking. I wisely did as she told me, and quit cold turkey. She reminds me that I quit after I was hypnotized, but I don't remember that part. I think it took almost a year for it to get it out of my system, and even then I didn't feel that I was as sharp mentally as I was when I was smoking.

The reason nicotine is so is addicting is because of the effect that it has on a least two of the bodies feel-good neurotransmitters, one of those is dopamine, and the other is the endorphins. The drug is so powerful that once someone gets addicted to it they stand only about a ten percent chance of beating the addiction. I have had patients with emphysema come to the office pulling an oxygen tank with a pack of cigarettes in their shirt pocket or pocket book. I have had patients with lung cancer continuing to smoke as if nothing was wrong. It is incredible how much denial the patient with nicotine addiction can have. The harmful effects of smoking are now well known: heart disease, hardening of the arteries, emphysema, lung cancer, increased risk of stroke and in women premature aging of the skin.

When I was a boy I used to help my neighbors put in tobacco in the summer time for five dollars a day. Almost everybody I knew back then smoked, including my father who died prematurely of high blood pressure and renal failure. Smoking is the worst thing that someone can do for their health and as a physician if I could I would ban tobacco, even though I know that wish is unrealistic.

Addiction to Marijuana and the other Hallucinogens

Another addiction that uses inhalation as a drug delivery system is the addiction to the active ingredient of marijuana, THC, tetra-hydro-cannabinols. It along with LSD, another hallucinogen, can also be addictive.

Alcoholism, Another Form of Endorphin Addiction

Another very common form of endorphin addiction comes from something that can be bought at any convenience store, namely alcohol. Alcohol itself is not an endorphin, but one of its effects is to cause the release of endogenous endorphins. Alcoholism is basically another form of endorphin addiction, and individuals with alcoholism probably have an endorphin system that is somewhat different from individuals who do not have alcoholism, and when someone with alcoholism drinks it causes a higher level of endorphins release than in non-alcoholics, and when they drink they don't just get drunk, they get high off of their own endorphins. When I have talked with patients who have alcoholism they confirm that this is indeed the way that alcohol makes them feel.

The person with alcoholism goes through the same steps that the person with heroin addiction does. As their alcoholism worsens they require more and more alcohol to get high, because over time their endorphin system requires more and more stimulation from alcohol to produce a level of endorphins that will produce a high, and as their drinking continues their endorphin factories begin to slack off on production as the body recognizes an excess of endorphins. The alcoholic now finds it harder and harder to reach the euphoric state associated with high endorphin levels. Individuals with chronic alcoholism are able to drink enormous amounts of alcohol because it takes increasing amounts of alcohol to produce even a basal endorphin level. If the individual with chronic alcoholism suddenly quits drinking they start going through endorphin withdrawal because their endorphin factories have lost their stimulation of alcohol. Severe alcohol withdrawal is called delirium tremens and can result in hallucinations, seizures, and death.

One of the facts that lends credence to the theory that alcoholism and narcotic addiction are both endorphin addictions is because naltrexone, an endorphin antagonist, can be used to treat both narcotic addiction and alcohol addiction.

Runner's High

Another type of endogenous endorphin addiction occurs in runners, and is called runner's high. Running probably causes the endorphin system to be stimulated to some degree in all runners, but in some runners apparently this effect is exaggerated, and when they run they develop euphoria from high level of their own endogenous endorphins. Some runners become addicted to this feeling and I have seen a number of patients who were addicted to running. It was very hard for me to get them to stop, even though several of them had serious medical conditions which made it dangerous for them to run, but some of them kept on running anyway, despite my admonitions to the contrary.

Addiction to Pain

Another form of endorphin addiction which to most of us would seem very bizarre is in the masochist, someone who is addicted to pain. I suspect in the masochist affliction of pain causes an exaggerated endorphin response, and like the runner they get addicted to high levels of their own endorphins. It would be interesting to see what effect naltrexone would have on the masochist. I am not aware of any such studies.

Addiction to the Manic Neurotransmitter

A very common addiction to another of the body's neurotransmitters is addiction to the manic neurotransmitter. The manic neurotransmitter is what gives us the confidence to be able to accomplish things and is the opposite of the depression neurotransmitter. We need a little squirt of it every morning to get going and have the confidence to get the day's task done, and when the amount of manic chemical matches the circumstance it's normal, but when someone has excessive levels of the manic neurotransmitter it causes them to feel overly confident, their brain to race, and make them feel invincible. They can lose all concern for the consequences of their behavior, have tremendous energy, go full stream for hours or days without sleep, have a hyperactive sex drive, and can spend money recklessly. One of the chief causes of excess manic neurotransmitter is the manic phase of untreated bipolar disease, when the manic center is pouring out high levels of manic neurotransmitter, but it can also occur when someone who is being treated for bipolar disease deliberately stops taking their medication in order to bring on a manic high. The supreme self-confidence that comes with high levels of endogenous manic neurotransmitter is very addictive, but an addiction that carries with it a very high price.

I recently saw a patient with bipolar disease in the office who had stopped her medication a month before and was in a full blown manic phase. This patient had been stable and doing well for several years without a single manic or depressive episode, but within a relatively short time after stopping her medication she was in a full blown manic phase. She was brought to the office by her family because they recognized that she was racing out of control. I had previously warned her family about this when she first got sick, so they were aware of the symptoms and knew something was not right. The patient herself didn't realize what was going on, which is not unusual, but fortunately her family did. She had to be hospitalized, and fortunately after a few days on medication she was back to normal.

Another cause of excessive manic neurotransmitter is through the stimulation of the manic center with drugs such as cocaine and ecstasy. We know that cocaine causes a rise in dopamine levels, but I believe that its primary effect is through the stimulation of the manic center, and dopamine is just one of the neurotransmitters through which the manic neurotransmitter has its effects. Cocaine is normally inhaled in the nose where it is absorbed into the blood stream through the mucous membranes, and from there it goes to the manic center where it has a powerful stimulating effect. Crack is a smokable form of cocaine which hits the manic center even quicker and with great intensity, but its effects are not as prolonged as inhaled cocaine. The popular party drug ecstasy also has its effects through the manic center. Though its effects are not as intense as cocaine, its effects are more prolonged. The belief on the streets is that ecstasy is not addicting, but in fact it can be just as addicting as cocaine.

The effect of cocaine and related drugs is to cause an individual to feel essentially the same as someone who is in the manic phase of bipolar disease. When I questioned one of my patients who has bipolar disease, and who has used cocaine, they confirmed that the feeling in both cases was essentially the same. The body rarely allows a free lunch, and the brain doesn't like it when it's homeostatic balance is upset. In many ways the brain and the body are like a sailboat. In a sailboat it is the weight in the keel that brings the boat back upright when the wind pushes the sails to one side, a homeostatic mechanism. The body has many homeostatic mechanisms that work to keep the body upright. In the case of endorphins, when the body recognizes an excessive amount of endorphins, it begins to shut down the endorphin factories. If the brain sees too much manic chemical, it will try to overcome the excessive manic chemical by producing its opposite, the depression chemical. When the stimulation of the manic center wears off the depression center doesn't shut down right away and continues to stay active for some period of time. This is what causes the rebound depression following the use of the stimulant drugs. Cocaine addiction is very malignant and the person who is addicted to cocaine usually will keep stimulating their manic center until their money runs out, because they know that when they come down from the high, depression will wash over them like a dark heavy smothering cloud.

Happiness Addiction

We usually don't think that there could be anything negative about happiness or gain, but for some individuals happiness can also be addicting. One of primary forms of happiness addiction is gambling. The person with gambling addiction becomes addicted to the big win or gain, and the feeling of happiness that goes with it, the bigger the win, the bigger the gain, the greater the happiness. Gambling in our society is endemic. We are surrounded by games and gambling of all kinds. People bet on cards, the slot machines, the lottery, horse racing, sports events, the stock market, credit default swaps, and on and on. People with severe gambling addiction sometimes lose everything in their quest for the next big win. They are certain that the next wager will give them the big win that they are looking for. As they get deeper and deeper in a hole they often risk more and more, until everything they own is gone. Up until that last bet they are sure that the next bet will result in the big win. Studies have shown that individuals with gambling addiction can be helped with the same types of twelve step programs as the other addictions.

Not too long ago I was talking with one of my sons, who is a college student, about addiction disease, and he told me about a new type of happiness addiction that has ensnared individuals of all ages, but especially young people. These are individuals who are addicted to video games that are played over the internet. Anyone with the software can join and become a part of the game. The game becomes like an alternate universe in which the players try to gain the next highest level in the game. The number of players in some of these games is in the millions, and they live all over the world, but especially in the Far East. The players take names and new identities in the games. Sometimes they play for hours and often stay up all night. Some of the players develop all of the signs of addiction and find it very difficult to stop. My son told me about one student that he had heard of, who in a moment of despair, when his character was defeated in one of the online battles, deleted his character, and committed suicide.

Another form of happiness-gain addiction is shop lifting. Frequently we read about a famous person who is arrested for shop lifting. They could easily buy what they steal, but they steal because they are addicted to the feeling of gain they have when they steal something and get away with it. Sooner or later they get caught, and often wind up paying a very high price for this addiction. Regular shopping can also be addictive. When we buy something new, it is normal to feel some degree of happiness, but some individuals become addicted to this feeling, and become compulsive buyers, most often for thing they don't need.

The most heinous form of the gain-happiness addiction is the individual who is addicted to rape and killing. I believe that what drives the serial killer is an addiction to the feelings of power and gain that they have over their victim, it's not the sex. In some ways this addiction resembles the stealing addiction, except in this case what is being stolen is the most precious thing a person possesses---their life. I remember hearing on NPR a jail house interview by a minister with the serial killer Ted Bundy. In the 1970s Ted Bundy raped and killed a number of coeds in Florida. He was intelligent and handsome, which made it easy for him to attract his victims. In the interview he talked about what he had done, and how he gone about it. I was struck by how much his description reminded me of an addiction. He described how he would look at pornography and fantasize about rape and murder. The more he thought about it the more intense his desire would become. Then one day he would go out rape and kill, and for awhile the cravings would subside, but then slowly the process would begin again, the intensity would build to a fever pitch, and finally he would go out, seek a victim, rape, and kill again. This went on for a number of years until he was finally caught, tried, and executed.

The serial rapist goes through the same process, fantasy, build up, and finally finding a victim and committing a rape. Again I don't think that it is the sex that the serial rapist is addicted to. The sex is only the instrument that the rapist uses to exercise his power, because the real addiction is not to the sex, but to the feeling of power, and gain over the victim.

Sexual Addiction

There is one form of addiction where the addiction is to the sex. I recently saw a TV documentary on PBS about sexual addiction, and the program pointed out that addiction to sex can occurs in both men and women from all walks of life, and from all levels of society. Sexual addiction has ruined countless lives, and in the news almost on a daily basis there is an article about someone of prominence who has gotten in trouble because of his or her sex life. [I initially wrote this before Tiger Wood's escapades became public, which really disappointed me personally because as golfer I had for many years admired him both as a golfer and individual]. Sometime ago I read about a candidate for the U.S. Senate who had to step down because it was revealed that he had pressured his wife, from whom he is now divorced, to go to a sex club where group sex was the attraction.

The person with sexual addiction is addicted to the high that comes from having sex with strangers, or some kind of new or unusual sex. Sexual addiction can occur in both heterosexuals and homosexuals. The individual with sexual addiction often seek out new partners and new ways of having sex in an effort to achieve a greater high from the next sexual experience. Their sex partners are often prostitutes who are more like objects to them than real people, and as impersonal to them as a drug. In a very real sense the individual with sex addiction also becomes like an object, as they allow their lower brain to take over their very being, and their every move becomes like a puppet on a string controlled by their lower brain. As their addiction intensifies their behavior often becomes more and more risky in an effort to satisfy their addiction, and the risk of exposure and the loss of their reputation becomes more and more likely. The man in the T.V. documentary was a minister, and except for a very understanding wife he would have lost everything.

Pornography also has the potential to become the vehicle for sexual addiction through fantasy and masturbation. One of the tragic consequences of addiction to pornography is that it can become so powerful that sex with a real person becomes meaningless, as fantasy becomes stronger than reality.

I believe that pedophilia is another form of sexual addiction. The recent scandal in the Catholic Church points out just how strong this addiction can be, because the priest who were guilty of abusing children were risking everything, including their very souls, in order to satisfy their addictive cravings. One of the things that has made this scandal so tragic is that the priests who were guilty of abusing children had been taught to know that it was wrong, and yet they allowed the craving chemicals in their lower brains to seduce their upper brains into doing things that their upper brains knew was wrong and would condemn them to Hell.

Thrill Addiction, the Addiction to Danger

Some individuals become addicted to the adrenalin or endorphin rush that accompanies danger, and will put themselves in situations where there is possibility of getting hurt or killed in order to feel that rush. One example of this is the sport of rock climbing, and just by coincidence, at the time I was writing this, I read where a world famous rock climber was killed attempting a very difficult climb on a sheer rock face. In the movie *The River Runs Through It,* the younger son, played by Brad Pitts, is addicted to thrill seeking, and despite the best efforts of his father and brother, he is eventually killed when his thrill addiction put him in harm's way one time too many.

Treatment of the Different Addictions

Addiction disease occurs when the power of the psyche shifts from the upper brain to the lower brain, and the lower brain is able to seduce the upper brain into a state of denial. Addiction disease can sometimes be very difficult to treat, but there are a number of treatments that can help. In addition to the twelve step programs, one way to treat addiction disease is to block the receptor site the person is seeking to fill. In endorphin addiction the endorphin receptor site can be blocked by using a narcotic antagonist like naltrexone, which keeps an individual from feeling the effects of an external endorphin like heroin. In alcohol addiction one approach is to use antabuse to keep the individual from drinking. If a person is taking antabuse and drinks, the antabuse will chemically combine with the alcohol to form formaldehyde and make them very sick. As long as they stay on antabuse they know they can't drink and this can help them stay sober. A substitute opioid agonist like methadone can be used to treat opioid addiction, because it can supply the body's basal endorphin level and eliminate the need for the addict to find a daily supply on the streets. In the case of cocaine addiction buproprion, an antidepressant, has been used, and in the case of sexual addiction Prozac has been used.

One of the key elements in the treatment of any addiction disease is building up the strength of the upper brain so it can say no to the cravings coming out of the lower brain and regain control of behavior. To use the horse and rider analogy, the rider must again learn to control the horse, when to make it walk or stop, and when to let it run. The programs which have had the most success have been the twelve step programs. In the twelve step programs, the first step is to get the person to admit that they have lost control of the horse. This first step is very important because denial is one of the hallmarks of addiction disease. The lower brain is very good at seducing the upper brain into getting what it wants. The person with addiction disease initially finds it very difficult to admit that the horse is running away, but once a person admits that they have lost control of the horse, the process of regaining control can begin. Regaining control of the horse basically means strengthening the upper brain to the point where it can so "no" to the cravings of the lower brain. Going to A. A., or N. A., is really a form of exercise to build up the strength of the upper brain, and going to A. A. for someone with alcoholism is like someone else going to the gym to keep their muscles and heart in shape, except in addiction disease it is the upper brain that needs strengthening. The upper brain needs to be strong enough so when the craving neurotransmitters of the lower brain come knocking, the upper brain can say "no".

The only criticism I have with the twelve step programs, particularly A. A., is they encourage people to label themselves by saying, "My name is .... and I am an alcoholic." I don't like labels because labels can get us into trouble. Maybe it's necessary in A.A., but when we label someone, or when someone labels themselves, that person can take on the characteristics of the label and may find it difficult to get rid of all of the characteristics of the label. Rather than saying, "I am an alcoholic," I think it is better to say, "I have alcoholism." When I have a patient in the office with alcoholism, or I think might be developing alcoholism, or who is at risk for developing alcoholism, I don't use the word alcoholic, because if you **have** something, it is much easier to deal with than if you **are** that thing.

Some Final Notes on Programming, Cognition, and Relationships

One of the primary causes of an incorrect emotional response is an error in the weighting of the significance of an event or circumstances, which is an error in cognition. Cognition can be influenced by any number of factors, but one factor often overlooked is the emotional climate that exists at the time the weighting takes place. Individuals with depression have a tendency to view circumstances *depressively*, and are apt to see the world from a helpless, hopeless point of view. Individuals with anger have a tendency to judge the world around them *angrily*, and individuals with excessive anxiety have a tendency to the see the world as a very scary place. How someone sees the world can be colored by the emotional glasses that someone is wearing at the time they are doing the weighing.

Another major cause for errors in the weighting process is the result of poor or negative programming that often occurs during childhood. If someone receives poor programming as a child, it will color how they see the world when they become an adult, unless somewhere along the line that programming is corrected. If a mother is overprotective, and never allows the child to spread his or her wings, or make mistakes, it is not surprising that when the child becomes an adult, he or she will likely see the world as a very scary place, and have anxiety problems.

Another programming error that parents make is when they equilibrate a child's behavior with their value. Some parents will tell a child when they behave well they are "good," and the child sees themselves as having value, but when the child behaves badly, which is inevitable, they are told that they are "bad," which causes their value to drop. Some children who are treated this way often wind up not being sure whether they are good or bad, whether they have value or are worthless. The best way to rear a child is to always give the child the feeling that they are loved and accepted, and have value all of the time; and it is only their behavior that is good or bad, and not the child themselves. I often see patients who are constantly striving for perfection, because they feel that they are loved and accepted only when they are perfect. They go through life feeling unloved and unaccepted, because they have never learned that it is not possible for we humans to attain perfection, and in their attempt to reach the goal of perfection, they often wind up with thoughts of helplessness and hopelessness, resulting in a chronic stimulation of the depression center, causing it to trickle out a steady stream of depression chemicals over a lifetime. This is probably one of the causes of a condition call dysthymia. If a child grows up in an environment which is very frightening, where the mother and father are constantly fighting, or threatening one another, or where one or both of the parents gets drunk and raise hell every night, it is only natural that the child will be programmed to see the world as a very scary place. A very high percentage of women I have seen in the office with anxiety and/or depression problems, give a history of childhood abuse that was mental, physical, or sexual, and sometimes all three. I believe that sexual abuse in females is far more common that is generally appreciated. Most of the time the abuse is by someone the child knows and trust: a relative, a stepparent, a neighbor. Childhood abuse of either a boy or girl, especially by a respected elder, is one of the most devastating and damaging things that can happen to a child. The recent scandal involving the abuse of children by priests, and the fact that it was tolerated for so long by some in the hierarchy of the Catholic Church, illustrates just how far society and the Church needs to go to eliminate and condemn an evil that permanently damages so many children.

Not all children who receive poor programming fail. Many children with poor programming go on to succeed and fulfill their potential through their own initiative, or through the help of others by getting rid of poor programming, and reprogramming themselves with a more realistic and optimistic view of themselves and the world.

For patients with generalize anxiety disorder, who seem to worry about everything, I sometimes offer the following guidelines to help someone get rid of some of their excessive worrying. *There are two things not to worry about, one is something that you can do something about, and the other is something you can't do anything about.* Patients with excessive worrying are encouraged to try and put their worries into one or the other of these two categories. If they can do something about a problem they are encouraged to do so. If they can't, they are encouraged to let it go, even though they may not like it. (This of course is a rephrasing of the serenity prayer made famous in A.A.) Many of the things that people worry about are things over which they have absolutely no control, and if they can eliminate some of their excessive worrying, it can make their lives much less stressful.

Interpersonal relationships are a major source of difficulty with couples they seem to fall into one of three patterns. The first is when one of the couple's is the master and the other the slave, with the master holding all of the power and the slave none. This usually makes the slave very resentful and can lead to feelings of helplessness and hopelessness. A second pattern is when they fight all of the time; one wins one time, and the other wins the next. Their relationship is a constant battle to see who can win. In this type of relationship the power base may be equal, but the aim of the couple is not to compromise or cooperate, but to see who can win. A third pattern is when the couples operate from an equal power base and have a sense of equality that allows them to solve problems through negotiation and compromise. I encourage couples who are having problems in their relationship to work toward the latter type of relationship. Put another way, if the husband wants to have more sex, he needs to wash more dishes.

Sometimes relationships between individuals don't work out, and very often there is much animosity and anger following the breakup. I urge couples to try and avoid the anger and animosity and encourage the couple to see the breakup as two gears that didn't match. When the gears don't match the machine won't run. But that doesn't mean that one gear is a good gear, and the other gear is a bad gear, or vice versa, it just means they don't match. When one of my sons marriage didn't work out, I encouraged him to take this attitude with his ex-wife. He and his ex-wife get along well and are raising a wonderful daughter, but I have seen far too many marriages where this was not the case, and the consequences are usually tragic for all concerned.

The Difference between Remorse and Guilt

When we hurt someone, (or if we use a religious frame of reference, sin against them), it is normal for most of us to feel bad about it. But is what we feel regret and remorse, or is it guilt? It makes a big difference. Guilt comes from an old English word *gylt*, which meant crime, sin, or fault. Regret comes from an old French word *regreter,* which meant to long after, bewail, or lament. Remorse also comes from a French word, *remors*, which meant deep regret. If it is guilt that someone feels, it usually means there is a need for some kind of punishment, or payment to atone for the sin. In regret, the individual will feel bad about what they have done, with a sense of remorse, but there is not the need for punishment. Sometimes in regret or remorse, we try to make up for a wrong, but it is not so much out of a sense of punishment, as it is out of a sense of making things right. The primary difference between guilt and remorse is what happens to the sense of self worth. In regret and remorse the self worth stays the same. We all goof up and make mistakes, we all have regrets, we are all human. When we goof up, or sin, it shouldn't mean that our sense of self worth has to drop. Guilt implies a need for punishment, and implies a diminution of self-worth. Jesus teaches us that we are all human, and if we sincerely ask for forgiveness, our sins, goof-ups, or mistakes will be forgiven, and our self worth can remain the same. The reason why this is so important is because I have seen so many patients over the years who feel guilty about some misdeed they have committed in the past which they have labeled unforgivable. They let their sins and mistakes brand them as being bad people, [that existential labeling thing again], and as a consequence their self worth drops. We all have done things we wish we had never done, but that doesn't mean we have to go through life trying to pay off a debt that can never be repaid. One of the reasons why I believe this is so important is because as an individual's self worth drops their ability to say no to their lower brain also drops, along with their self control, self esteem, and potential. I have found that the individuals who are the most ridden with guilt are also the ones who have the least control over their lower brain, and the ones most likely to get into trouble with the law and relationships. This is not to say that parents or society shouldn't have the right to punish individuals who disobey either the parent's or society's rules, but there is a difference between the individual punishing themselves for some sin they feel is unforgiveable; and society or parents punishing them because they have disobeyed the rules or a law..

One Final Note

In 1960 Dr. Thomas Szasz wrote a book entitled *The Myth of Mental Illness*. The thesis of Dr. Szasz's book was that there was no such thing as mental illness and it belonged in the same category as witchcraft. In a prelude to the book he wrote the following:

"...the notion of mental illness thus serves mainly to obscure the everyday act that life for most people is a continuous struggle, not for biological survival, but for " a place in the sun," "peace of mind," or some other human value.

And he went on to say:

"Sustained adherence to the myth of mental illness allows people to avoid facing this problem, believing that mental health, conceived as the absence of mental illness, automatically insures the making of right and safe choices in one's conduct of life. But the facts are the other way. It is the making of good choices in life that others regard, retrospectively, as good mental health."

The project of this book is to refute this argument and make the case that it is not necessarily the circumstances, the choices, or behaviors that determine why someone feels or behaves the way they do. I have seen thousands of patients whose emotions failed to match the circumstances in their lives, and it wasn't because they made a choice about their emotions or life circumstances. These were individuals with bipolar disease, panic attacks, major depression, and PTSD, individuals who had emotions they did not choose, and over which they had absolutely no control.

If our emotions were music, the tone of the music, the key, the melody, the timbre, should match the circumstances. In happy times the music should be light and gay, and in a major key; in sad times, dark and slow and in a minor key. But in some people's lives the music doesn't come out right. Sometimes it is because they fail to read the notes correctly. And sometimes it is because they play the wrong instrument, they play the basses when they should be playing the flutes and the piccolos. But with help they can learn to play the right instrument and read the music correctly, and when they do, the music of their lives will come out right.

But there is another reason why the music doesn't come out right, sounds dissonant, and off key. Imagine the brasses start blaring in the middle of a lullaby, or the basses keep up a mournful grind when the music should be light and gay, and the conductor can't get them to stop. It's not always because someone can't read the notes, or plays the wrong instrument, sometimes it's because a section of the orchestra has a will of its own. And that's what it's like to have bipolar disease and panic attacks, when one or more of the primary survival emotional centers has a will of its own, and the individual can't make it stop. And that is why Dr Szasz is wrong.

What does it mean to be healthy? The words heal and healthy come from the same root word *healen*. In Old English the word meant to *make* whole or sound, and if we use the original meaning of the word, being healthy is not a state into which we are born, instead it is a state that we achieve through healing. When we say that someone is physically healthy, it means that their bodily functions fall within a normal range. The TSH which is a measure of thyroid function, and stands for thyroid stimulating hormone, should be between 0.35 and 4.35. The blood sugar should be between 80 and 130, depending on whether someone has just eaten. When all of the physical parameters of the body fall within normal limits, we say that an individual is healthy. But if the TSH is 30, or the blood sugar is 500, these readings are outside the normal limits and it means that a disease state exists. [The word disease originated from two words that describe exactly what they mean, *dis*-without or a lack of, and *aise* or ease… a lack of ease. It entered the English language sometime before 1338.] In the case of the TSH, when it is 30, it means that the thyroid is under functioning and the brain is calling for more thyroid hormone, a condition called hypothyroidism. When the blood sugar is 500, it means that the pancreas gland is not putting out enough insulin and the patient has diabetes. We can correct the thyroid disease by giving thyroid hormone and the diabetes by giving insulin, and when we do we make the patient whole and healthy.

But what about our mental and emotional health? What criteria can we apply to that aspect of our lives? I see no reason not to apply the same criteria to the mental and emotional part of lives that we apply to the physical part of our lives. When an individual is mentally and emotionally healthy their emotions should fall within a normal range and match the circumstances. It is not normal for someone to have a level ten of anxiety when they are sitting in church surrounded by family and fellow worshipers. Nor is it normal for someone to have depression when everything in their life is going well. With medication or psychotherapy, and sometimes both, it is possible to bring someone's emotions back into a normal range, and then we can say that person is mentally and emotionally healthy. Why should we be any less inclined to treat excessive anxiety or depression, than we do to treat high blood pressure or high cholesterol, because both are disease states and both deserved to be treated.

I believe that education is one of the keys. I don't believe that anyone should be forced to receive treatment, unless they are truly psychotic and even then the circumstances need to extraordinary, but I do believe that when most individuals are educated about their emotions, and understand what makes them abnormal, most will chose to get help for their mental and emotional problems. Some have questioned whether abnormalities of the emotions should be treated. Dr. Peter Kramer, a psychiatrist, wrote a book several years ago called *Listening to Prozac*. The theme of his book was a concern that Prozac, by getting rid of people's depression, was making them feels too good, better than normal; and maybe depression shouldn't be treated. But he never really said what normal was, or how someone should feel. In a similar vein some doctors and laypersons have questioned whether it is appropriate to give someone with severe anxiety benzodiazepines. Their concern, like Dr. Szasz, is that individuals with anxiety suffer from a weak will, or are reacting to the normal stresses of everyday life, and should be able to pull themselves up by their boot straps. Anyone who feels this way should take a lesson from Dr. Kramer, who after thinking about it for a while, came back and wrote another book, *Against Depression*, in which he acknowledged that depression is a condition which should be treated, and that society and medicine needed to recognize it for what it is, a disease, and give it the same respect afforded heart disease and cancer.

When I see a patient in the office with an emotional problem, one of the first things that I try to do is to educate the patient about what my goals in therapy are. I explain to them that I'm not going to try and make them feel better than normal, or happy all the time. I explain to them that no amount of anti-depressants can take away the pain of losing a loved one, or an anti-anxiety agent take away the fear from a wife whose husband is beating her. My goal in therapy is to try and help the patient bring their emotions back into a normal range, one that matches the circumstances. If their emotions fall outside the normal range, I will try to help them figure out why, but I will need their help, because I can't fix their problem alone. I explain to them that none of us gets it right all of the time, but it is possible for most of us to get it close, most of the time. If it is their programming, or cognition, that is off, I will help them correct their faulty programming, or send them to someone who can help through psychotherapy. If one of their primary survival emotional centers is not working properly, more than likely they will need medication. In some cases it will require both psychotherapy and medication.

I try to forewarn patients that one of the hazards they face when their emotions change and become more normal, is it won't feel normal to them, because the way they felt before therapy may feel more normal to them than the way they feel after therapy. I try to forewarn them as their emotions change at first it will feel strange. The tragedy of this dilemma is that all too often patients give up on medication and a new way of life, or a new way of thinking, and slip back into old habits. Old ways of feeling, thinking, and behaving die hard. So both the therapist and the patient should beware.

Finally, I would hope that we could all learn to look at the emotions the same way that we look at the blood sugar and thyroid hormone, and have the same attitude toward both. Being healthy and free of disease in the physical, and the mental-emotional part of our lives, means the measurements in both should fall within a normal range. My hope is that one day society will place the same emphasis on the mental-emotional part of our lives as they do on the physical. To do anything less is a travesty.

William W. Hedrick, M. D.

Post Script:

I am aware that the paradigm I have presented runs counter to much of what is currently being taught and will likely meet with much resistance and skepticism by the psychiatric and psychological community. I only ask that my ideas be given a fair reading and not be dismissed out of hand. The hypothesis that I have presented explains far more of what is seen in the real world than the one currently in vogue, and I challenge anyone to show evidence to the contrary. How else can you explain how the stimulation of area Cg 25 can turn off depression, or how someone with bipolar disease can feel both manic and depressed at the same time, the so called mixed state, or how someone can have spontaneous bouts of manic, depression, anger and anxiety, a condition I have labeled quadripolar disease, unless, as I have proposed, there are specific emotional centers in the brain that put out the neurotransmitters that cause someone to feel those emotions. I believe the current research is being aimed at the secondary players in the mental health game, and the real players are in the primary survival emotional centers where the neurotransmitters of anxiety, anger, manic and depression originate, and I would urge the research be directed more toward the identification of those neurotransmitters.

I believe that one day in the future Dr. Freud's prediction will come true that the answers to mental illness will be found in the laboratory, and when that day comes psychiatry and psychology will truly become a science. My prediction is that the paradigm that I have laid out will be the foundations of that science.

WWH

References

1. The Mind, Richard M. Restak, M.D., Bantam Books, 1988

2. Molecules of Emotion, Candace B. Pert, Ph.D. Touchstone Book, 1997

3. The Selling of DSM, The Rhetoric of Science in Psychiatry, Stuart A. Kirk and Herb Kutchins, Aldine De Gruyter,

4. A New Guide to Rational Living, Albert Ellis, Ph.D. and Robert Harper, Ph.D., Prentice Hall, Inc. 1975, 1961

5. Critical Condition, How Health Care In America Became Big Business-And Bad Medicine, Donald L. Barlett and James B. Steele, Doubleday, 2004

6. A History of Psychiatry, From the Era of the Asylum to the AGE of Prozac, Edward Shorter, Doubleday, 1997

7. Looking for Spinoza, Joy, Sorrow and the Feeling Brain, Antonio Damasio, M.D. Harcourt, Inc. 2003

8. Madness in America, Cultural and Medical Perception of Mental Illness before 1914, Lynn Gamwell and Nancy Tomes, Cornell University Press, 1995

9. Mysteries of the Mind, Richard Restak, M.D., National Geographic 2000

10. DSM-IV, Diagnostic and Statistical Manual of Mental Disorders, Published by the American Psychiatric Association, 1994

11. When I Say No, I Feel Guilty, Manuel J. Smith, Ph.D. Bantam Books, 1973

12. Emotions and Life, Perspectives From Psychology, Biology, and Evolution, Robert Plutchik, Ph.D., American Psychological Association, 2003

13. The Wisdom of the Body, Walter B. Cannon, M.D., W.W. Norton and Company, Inc. 1932, 1939

14. The Story of Psychology, Morton Hunt, Doubleday, 1993

15. The Barnhart Concise Dictionary of Etymology, Edited by Robert K. Barnhart, Harper Collins Publishers, 1995

16. Bullying Behaviors Among U.S. Youth, Tonja Nansel, and et.al. JAMA, April 25, 2001.

17. Tetrahydro-B-Carboline Micro-Injected in the Hippocampus Induces an Anxiety-Like State in the Rat, Pirkko Huttunen and R.D. Myers, Pharmacology Biochemistry and Behavior, Vol. 24, 1986

18. Compulsive Gambling, The Main Event, February, 1987

19. The Mind of the Rapist, Newsweek, July 23, 1990

20. Possible Antidepressive Effects of Opioids: Action of Buprenorphine, H. M. Emrich, P. Vogt, and A. Hrez, Max-Planck Institute of Psychiatry Munich, Germany

21. Hooked, Why Isn't Everyone an Addict, Deborah Franklin, Hippocrates, Nov. Dec. 1980

22. Making Us Crazy, Herb Kutchins and Stuart A. Kirk, The Free Press, Nov. 1997

23. Regional Disturbances, Lawrence Osborne, New York Times, May 6, 2001

24. Woman found 8 days after crash, Associated Press, Sept. 29, 2007

25. Effect of TV Violence on Sexual Arousal in Men, Neil M. Malamuth, Ph.D., Medical Aspects of Sexuality, Oct. 1987

26. Buprenorphine Treatment of Refractory Depression, J. Alexander Bodkin, M.D., Gwen L. Zornberg, M.D. et.al, Journal of Clinical Psyhopharmology, 1995

27. Methadone and Opiate Drugs: Psychotropic Effect and Self-Medication, Gerald McKenna, Annals New York Academy of Sciences, 1982

28. The Ancient Roots of Modern Melancholy, Did Depression Help Our Ancestors Survive? Josh Fischman, U.S. News and World Report, Feb. 14, 2000

29. Teen, Thought Dead, Found Alive in Ravine, Melanthia Mitchell, Associated Press, Oct. 12, 2004

30. Did Antidepressants Depress Japan? Kathryn Schuiz, New York Times Magazine, Aug. 22, 2004

31. Opioid Peptides in Affective Disorders, Philip A. Berger and Charles B. Nemeroff, Psychopharmacology, The Third Generation of Progress, 1987

32. Pharmacological Challenges to the Endogenous Opioid System in Affective Illness, Martin R. Cohen, M.D., and David Pickar, M.D., The Journal of Clinical Psychopharmacology, Vol. 1, No. 4, 1981

33. Involvement of Opioid Receptors in the Effects Induced by Endogenous Enkephalins on Learned Helplessness Model, Purificacion Tejedor-Real, et.al, European Journal of Pharmacology, 354, 1998

34. Isolation and Partial Characterization of an Opioid-like 88 kDa Hibernation-Related Protein, Noel D. Horton, Peter R. Oeltgen, et.al, Comparative Biochemistry and Physiology, 1998

35. A Possible Opioid Receptor Dysfunction in Some Depressive Disorders, Irl Extein, Annals New York Academy of Science, 1982

36. Seasonal Affective Disorder, Hibernation, and Annual Cycles in Animals: Chipmunks in the Sky, N. Mrosovsky, Journal of Biological Rhythms, Vol. 3, No. 2 1988

37. Sexual Effects of Movie and TV Violence, George A. Comstock, Ph.D., Medical Aspects of Human Sexuality, July 1986

38. Naltrexone in the Treatment of Alcohol Dependence, Joseph R. Volpicelli, M.D. et.al, Archives of Gen. Psychiatry, Nov. 1992

39. Scientific Medicine in the XIX Century, The Genesis of Claude Bernard, pg. 279-281

40. Cocaine Abuse: Relapse-Prevention Pharmacotherapies, Thomas R. Kosten, M.D., Clinical Advances, 1993

41. Insights About Pathological Gamblers, 'Chasing Losses' in Spite of Consequences, Capt. John R. Cusack, et.al, Postgraduate Medicine, April 5, 1993

42. Sex Addictions, Carolyn F. Cordasco, EdM, CCSW, NCMJ, Sept. 1993

43. Should Compulsive Sex Be Considered an Addiction? Family Practice News, March 15, 1992

44. Enhanced Sensitivity of Pituitary B-Endorphin to Ethanol in Subjects at High Risk of Alcoholism, Christina Gianoulakis, Ph.D., et.al, Archives of General Psychiatry, Mar. 1996

45. Bereavement Counseling: The Physician's Role, Charles Driscoll, M.D., The Female Patient, March, 1989

46. The Relationship Between Anger and Depression, Maurzio Fava, M.D. and Jerrold F. Rosenbaum, M.D., Clinical Advances in the Treatment of Psychiatric Disorders, April, 1993

47. The Question of Pornography, It Is Not Sex, But Violence, That Is An Obscenity In Our Society, Edward I. Donnerstein and Daniel G. Linz, Psychology Today, Dec. 1986

48. Fertile Minds, From birth, a baby's brain cells proliferate wildly, making connections that may shape a lifetime of experience. The first three years are critical. J. Madeleine Nash, Time Magazine, Feb. 3, 1997

49. A Review of Unrecognized Mental Illness in Primary Care, Edmund S. Higgins, M.D. Archives of Family Medicine, Oct. 1994

50. Psychiatry's Global Challenge, Arthur Kleinman and Alex Cohen, Scientific American, March, 1997

51. Making Love, Not War, Alfie Kohn, Psychology Today, June, 1988

52. Toward A Psychology of Evil, A. Scott Peck, Simon and Schuster, 1983

53. The Myth of Mental Illness, Dr. Thomas Szasz, American Psychologist, 15,113-118 1960

54. Mystical Bedlam, Madness, Anxiety, and Healing in Seventeenth Century England, Michael MacDonald, Cambridge University Press, 1981

55. The Nature of Melancholy, From Aristotle to Kristeva, Jennifer Radden, Oxford University Press, 2000

56. The Loss of Sadness, How Psychiatry Transformed Normal Sorrow Into Depressive Disorder, Allan Horwitz and Jerome Wakefield, Oxford University Press, 2007.

www.ingramcontent.com/pod-product-compliance
Lightning Source LLC
Chambersburg PA
CBHW022056210326
41519CB00054B/484